Circular Economy and Efficient Use of Resources

Circular Economy and Efficient Use of Resources

Editor

Lucian-Ionel Cioca

MDPI • Basel • Beijing • Wuhan • Barcelona • Belgrade • Manchester • Tokyo • Cluj • Tianjin

Editor
Lucian-Ionel Cioca
Industrial Engineering and
Management Department,
Faculty of Engineering
Lucian Blaga University of Sibiu
Sibiu
Romania

Editorial Office
MDPI
St. Alban-Anlage 66
4052 Basel, Switzerland

This is a reprint of articles from the Special Issue published online in the open access journal *Processes* (ISSN 2227-9717) (available at: www.mdpi.com/journal/processes/special_issues/ economy_resources).

For citation purposes, cite each article independently as indicated on the article page online and as indicated below:

LastName, A.A.; LastName, B.B.; LastName, C.C. Article Title. *Journal Name* **Year**, *Volume Number*, Page Range.

ISBN 978-3-0365-2994-3 (Hbk)
ISBN 978-3-0365-2995-0 (PDF)

© 2022 by the authors. Articles in this book are Open Access and distributed under the Creative Commons Attribution (CC BY) license, which allows users to download, copy and build upon published articles, as long as the author and publisher are properly credited, which ensures maximum dissemination and a wider impact of our publications.

The book as a whole is distributed by MDPI under the terms and conditions of the Creative Commons license CC BY-NC-ND.

Contents

About the Editor vii

Preface to "Circular Economy and Efficient Use of Resources" ix

Catalin Popescu and Manuela Rozalia Gabor
Quantitative Analysis Regarding the Incidents to the Pipelines of Petroleum Products for an Efficient Use of the Specific Transportation Infrastructure
Reprinted from: *Processes* **2021**, *9*, 1535, doi:10.3390/pr9091535 1

Hafezali Iqbal Hussain, Muhammad Haseeb, Fakarudin Kamarudin, Zdzisława Dacko-Pikiewicz and Katarzyna Szczepańska-Woszczyna
The Role of Globalization, Economic Growth and Natural Resources on the Ecological Footprint in Thailand: Evidence from Nonlinear Causal Estimations
Reprinted from: *Processes* **2021**, *9*, 1103, doi:10.3390/pr9071103 25

Noor Hafidzah Jabarullah, Afiqah Samsul Kamal and Rapidah Othman
A Modification of Palm Waste Lignocellulosic Materials into Biographite Using Iron and Nickel Catalyst
Reprinted from: *Processes* **2021**, *9*, 1079, doi:10.3390/pr9061079 39

Muhammad Mohsin, Qiang Zhu, Sobia Naseem, Muddassar Sarfraz and Larisa Ivascu
Mining Industry Impact on Environmental Sustainability, Economic Growth, Social Interaction, and Public Health: An Application of Semi-Quantitative Mathematical Approach
Reprinted from: *Processes* **2021**, *9*, 972, doi:10.3390/pr9060972 51

Amalia Furdui, Lucian Lupu-Dima and Eduard Edelhauser
Implications of Entrepreneurial Intentions of Romanian Secondary Education Students, over the Romanian Business Market Development
Reprinted from: *Processes* **2021**, *9*, 665, doi:10.3390/pr9040665 71

Miorita Ungureanu, Juhasz Jozsef, Valeria Mirela Brezoczki, Peter Monka and Nicolae Stelian Ungureanu
Research Regarding the Energy Recovery from Municipal Solid Waste in Maramures County Using Incineration
Reprinted from: *Processes* **2021**, *9*, 514, doi:10.3390/pr9030514 103

About the Editor

Lucian-Ionel Cioca

Lucian-Ionel CIOCA is a full professor at the Lucian Blaga University of Sibiu, which he acquired in 2007. Since 2010, he has been a doctoral advisor in Engineering and Management. His research focuses on human resources management, production systems engineering, ergonomics, circular economy, sustainability, occupational safety, and health management. He has published over 250 scientific papers, of which Clarivate indexes over 135 in the Web of Science. He is a member of the following professional organizations: the German General Engineer's Association, the World Economics Association (WEA), the United Kingdom, the Romanian Academy of Scientists. In recognition of his research activity, he is the Editor-in-Chief of *INMATEH - Agricultural Engineering*, Associate Editor of *Frontiers* in Psychology, and Topic Editor of *Frontiers* in Environmental Science; journals indexed *Web of Science and Scopus*. He is also an editorial board member of the following journals indexed in *Web of Science and Scopus*: *International Journal of Environmental Research and Public Health, Polish Journal of Management Studies, Quality-Access to Success, Sustainability, and Safety*. He is a guest editor of the following journals indexed in *Web of Science and Scopus*: Advances in Materials Science and Engineering, *Frontiers* in Psychology, International Journal of Environmental Research and Public Health, Processes, Sustainability, and Safety. Professor Lucian-Ionel CIOCA is a reviewer of scientific articles published in journals indexed in Web of Science and Scopus, for example, Studies in Educational Evaluation, IEEE Transactions on Industrial Informatics, Studies in Informatics and Control; Waste Management, Journal of Hazardous Materials, Water Environment Research, Symmetry, Resources, Sustainability, and Safety.

Preface to "Circular Economy and Efficient Use of Resources"

Previous and current resource use models have led to high levels of pollution, environmental degradation, and depletion of natural resources. The waste policy has a long history and has traditionally focused on more environmentally sustainable waste management. To be efficient in terms of resource use, the principles of the circular economy should change this trend, transforming the economy into a sustainable economy. The new waste regulations introduce new waste management objectives in terms of waste prevention, reuse, recycling, and storage by addressing new processes.

Lucian-Ionel Cioca
Editor

Article

Quantitative Analysis Regarding the Incidents to the Pipelines of Petroleum Products for an Efficient Use of the Specific Transportation Infrastructure

Catalin Popescu [1,*] and Manuela Rozalia Gabor [2]

1. Department of Business Administration, Petroleum-Gas University of Ploiesti, 100680 Ploieşti, Romania
2. Department ED1-Economic Sciences, "G. E. Palade" University of Medicine, Pharmacy, Science and Technology of Târgu Mureș, 540142 Târgu Mureș, Romania; manuela.gabor@umfst.ro
* Correspondence: cpopescu@upg-ploiesti.ro

Abstract: The transportation infrastructure for petroleum products contains complex pipeline systems, developed on a global scale and totaling investments of hundreds of millions of dollars. The operation and maintenance of these systems have to be performed in relation to the analysis of incidents of various types, which take place in various areas of the world. The present paper aims to analyze in as much detail as possible, from a statistical point of view, the case of the pipeline system for petroleum products in Romania in order to streamline the operation of this critical infrastructure for Romania. Through the statistical tools, we established the hierarchies of the causes of the analyzed incidents, weights of the effects generated by these sources of accidents, and correlations between various parameters, in order to create a useful plan of measures and actions in the efficient operation of the pipeline system. The importance and topicality of the subject is also demonstrated by the major negative impact of the accidents in this sector, through product leaks from pipes in the soil and in watercourses, which generate significant pollution values, thus influencing the balance of the environment.

Keywords: oil pipeline; incidents; cause; petroleum products; statistical analysis; consequences; safety; Chi-Square test; cross-tabulation

Citation: Popescu, C.; Gabor, M.R. Quantitative Analysis Regarding the Incidents to the Pipelines of Petroleum Products for an Efficient Use of the Specific Transportation Infrastructure. *Processes* 2021, 9, 1535. https://doi.org/10.3390/pr9091535

Academic Editor: Lucian-Ionel Cioca

Received: 2 August 2021
Accepted: 25 August 2021
Published: 28 August 2021

Publisher's Note: MDPI stays neutral with regard to jurisdictional claims in published maps and institutional affiliations.

Copyright: © 2021 by the authors. Licensee MDPI, Basel, Switzerland. This article is an open access article distributed under the terms and conditions of the Creative Commons Attribution (CC BY) license (https://creativecommons.org/licenses/by/4.0/).

1. Introduction

The world's oil and gas pipeline system covers hundreds and thousands of miles. This has been conducted as a major investment for areas or countries that do not have such resources to benefit from their processing. These investments have now become a priority for companies that own these pipelines, as incidents have begun to occur and it is necessary to monitor and manage such situations. Incidents such as accidents, breakdowns, or failures are unfortunate events because of the consequences they entail: in some cases, the consequences can be economic, environmental or, in the worst conditions imaginable, accidents that can cause loss of life [1]. Pipeline safety and integrity are crucial for a sustainable future and responsible development [2]. Precisely out of the desire to ensure increased safety in the transport of petroleum products, it is necessary to analyze in as much detail as possible the causes of incidents produced over time.

Basically, the main question for this study is: what are the main causes in the generation of incidents in the pipeline system of petroleum products? This must be ascertained in order to design appropriate measures and actions, including maintenance solutions, to make the specific transport infrastructure more efficient and less polluting.

Therefore, this study identifies the main factors and causes of incidents for the pipeline system of petroleum products in Romania. Available data from 2017 to 2019 are statistically analyzed. There are generated hierarchies for causes of incidents, and correlations are checked for different parameters, related to the pipeline incidents. The analysis is necessary for the implementation of a plan of measures to include: investments in equipment and for the replacement of some sections of pipes that have been affected; protection of lands that have pipes in their basement; complex measures for monitoring areas that have pipelines; updated maintenance plans, etc.

The causes of oil spills must be known, analyzed, and treated in order to eliminate the loss of oil products through pipeline systems and protect the environment.

There are various databases around the world related to pipeline incidents in the transportation of petroleum products, as follows:

- In the US, Pipeline and Hazardous Materials Safety Administration (PHMSA);
- In Canada, Pipeline Incident Database (PID);
- In the United Kingdom, United Kingdom Onshore Pipeline Operators' Associations (UKOPA);
- In Europe, European Gas Pipeline Incident Data Group (EGIG);
- In Russia, initially National Technical Inspectorate and then Federal Service for Environmental, Technological and Nuclear Supervision; and
- In Australia, Australian Pipeline Industry Association (APIA).

Most countries in the world (including Romania) do not have a database system for reporting oil and gas pipeline incidents. Why would a globally unified database be needed? Because each database, at regional or national level, contains different criteria for reporting incidents in this category. In addition, the presentation and debate of cases declared at the level of certain areas or countries must be conducted through the prism of common, standardized elements and must be unanimously accepted by experts.

In addition, there are organizations and associations that specialize in conducting studies dedicated to this sector. One such globally recognized and representative entity is the European Oil Company Organisation for Environment, Health and Safety (CONCAWE). CONCAWE, a European association that includes a group of leading oil companies (more than 40), carries out regular research on environmental issues relevant to the oil industry. The topics cover wide areas, such as: fuel quality and emissions, air quality, water quality, waste, soil contamination, cross-country pipeline performance, etc.

At the same time, some specialists describe, in a simplified way, the causes of pipeline failure. For example, a classification was proposed with four sources of incidents [3]:

- Third-party damage;
- Corrosion;
- Design and construction error; and
- Incorrect operation conditions.

In order to demonstrate the lack of unity of points of view in classifying the causes of pipeline incidents, two of the most representative databases are presented: PHMSA and EGIG.

PHMSA database proposes a system that contains eight categories of pipeline failure causes: corrosion (external; internal; stress corrosion cracking; selective seam corrosion); excavation damage; natural force damage; material/weld failure; equipment failure; incorrect operation; and all other causes. EGIG database has a classification that contains only five categories: corrosion; external interference; construction defect/material failure; ground movements; other and unknown.

In the US, pipeline operators are required by law to report pipeline incidents, while in Europe this is not mandatory.

The importance of the subject is demonstrated by the fact that such accidents incur high material costs for the oil pipeline's operating companies and significant damage to the environment, people, and property in the vicinity of the pipeline failures.

The topicality of the studied topic is proven by the provision of information based on the content of the PHMSA database in the period 2010–2020 (Table 1). From these data, it is easy to deduce the major negative effect produced by these incidents from the point of view of the affected persons, on the environment and from a financial point of view.

Table 1. Summary of pipeline incidents in the US from 2010 to 2020 (authors' own processing from PHMSA database).

Report	Incidents	Injuries	Fatalities	Evacuees	Fires	Explosions	Damages ($)
Gas Distribution	1222	539	115	27,870	714	273	2,408,976,046
Gas Transmission & Gathering	1369	110	27	13,153	146	63	1,492,746,535
Hazardous Liquids	4359	39	14	2780	145	19	2,962,900,530
Totals	6950	688	156	43,803	1005	355	6,864,623,111

On the other hand, at present, the Romanian national company operates a pipeline transport system with a length of 3809 km, of which 3161 km (82% of the total) is actually used for the transport of crude oil, gasoline, condensate, and liquid ethane. The action area is located mainly in the southern part of the country and with a direct connection to the main port on the Black Sea, Constanta.

The crude oil transport via pipelines in Romania has a history of over 115 years. In 1901, the first crude oil transport via pipelines in Romania was along the route Buștenari-Băicoi Rail Station, Prahova County. Today, the company transports crude oil via the national pipeline system describing 3800 km in length and 27 million tons' throughput, crossing 24 counties. The maximum allowable losses during transportation are <0.365% from the total transported quantity; otherwise, the company should pay taxes due to the losses incurred and environmental pollution.

Therefore, the crude oil transport activity must be carefully monitored so that the number of incidents in the pipeline system decreases and the negative impact, generated by these incidents, manifests itself on a much smaller scale.

The paper is designed in a standard way, so that after the Introduction, Section 2 is dedicated to Literature Review, then Section 3, entitled Materials and Methods, is integrated, followed by Section 4 for Results, and finally, Section 5, containing Conclusions, is included.

2. Literature Review

The pipelines are considered the safest way to transport petroleum products [4]. Actually, the idea of using pipelines to transport hazardous products as a preferred method is related to the safety and cost, when compared to train and ground transportation [5]. Still, the pipeline systems are associated with risks, leading to negative consequences [6,7].

Oil spills are environmental disasters and their long-term impact is not just a concern for the environment and economy, but also for health and well-being of all living things [8]. Moreover, along with the benefits of pipelines come the risks to health and property generated by fires and explosions [9].

The study of accident causes for the pipeline industry is relatively rare, which severely restricts the perception of pipeline accidents and limits the adequacy and timeliness of the proposed response measures [10] (p. 1).

In principle, CONCAWE produces extensive studies on most of the topics mentioned above annually or every few years. The most recent study on the subject of the oil pipeline is from May 2021 and considers the performance of European cross-country oil pipelines. Specifically, this report covers an important period, namely 1971–2019, in connection with spillage data on European cross-country oil pipelines, referring to a current network of pipelines in Europe of approximately 36,000 km, which provides annual transport of 620 million m^3 of crude oil and petroleum products. Below, the next section will include some useful information from this report regarding spillage causes and other incident causes for the oil pipelines.

In the last 10–15 years, some specific analyses regarding the pipeline incidents were carried out. In 2013, a study regarding transportation of gas and hazardous liquid, carried out by representatives from the Manhattan Institute, stated that road transportation had an annual accident rate of 19.95 incidents per billion ton miles, while rail transportation had 2.08 incidents per billion ton miles, compared to 0.89 incidents per billion ton miles for natural gas transmission, and 0.58 incidents per billion ton miles for hazardous liquid pipelines [11].

In 2015, a statistical analysis was published which declared that over 75% of accidents were caused by third party interference, external corrosion, material failure, and internal corrosion in the case of the onshore gas transmission pipelines in the US [12].

One year later, the pipeline incident data for the onshore gas transmission pipelines in the US were investigated, and it was stated that 53% of the accidents occurred on pipelines installed between 1950 and 1960 [13].

Many specialists consider the issue of the mechanical integrity and ageing of the pipeline systems to be critical. In this sense, it is known that most European pipeline systems were built in the 1960s and 1970s. In 2019, less than 2% of the pipelines were 10 years old or less and 70% were over 40 years old [14]. In the same time, 40% of the pipeline networks worldwide have reached their projected 20-year service lifetime [15].

There are also analyses/studies that claim that complex enviro-technical systems, such as oil pipelines which are characterized by oil spills, are designed, firstly, for economic efficiency rather than environmental protection [16].

In the face of accidents, people always think about the causes of accidents. This way of thinking led to the theory of accident causes and became the theoretical basis for understanding accidents [17].

According to different statistics reports regarding the pipeline accidents [18,19], the causes of pipeline accidents are manifold, caused by multiple factors.

The views expressed in this regard are diverse. For example, a first view states that the spillage causes can be grouped into five main categories: mechanical failure, operational, corrosion, natural hazard, and third party [14].

Another approach considers that the pipe incidents appear where corrosion, degradation, inadequate installation, or manufacturing defects affect the pipes' structural integrity [20].

Accidents that cause product spills have even more dangerous consequences if they occur near to the ignition sources and under certain conditions [21]. As a confirmation, another research study identified that the most dangerous scenarios are oil spills, fire, and oil vapor explosion due to the loss of piping integrity (rupture) of the pipeline's section [22].

At the same time, these pipe accidents generate important economic losses every year and include property damage, commodity loss, and/or environmental remediation [23]. Therefore, the costs produced by a loss of containment are used for risk-based decision-making processes [24]. Additionally, the costs are often used to classify the severity of pipeline failures [25]. On the other hand, risk-based decision-making processes have as a defining tool a risk assessment approach. In this regard, a risk assessment is carried out by estimating the probability of occurrence and the severity of the consequences that this event may produce [26].

In order to reduce the risk of leakage accidents and to prevent major spills, it is necessary to conduct safety assessments of heavy oil gathering pipelines. In many situations, failure data for these pipelines are insufficient or irrelevant, and the use of statistical methods is difficult, so a risk assessment system for heavy oil gathering pipelines is proposed in the absence of failure data [27]. To estimate the risk of oil pipeline failure, different risk assessment methods are used; for example, event tree analysis [28,29], fault tree analysis [28,30], bowtie [31,32], and others [28,32].

Given the complexity of pipeline transportation of petroleum products, risk management strategies should no longer be selected solely in terms of economic and technical aspects. Decision makers have to address the sustainability of risk management by assessing the effect of their decisions regarding the sustainable development of a given territory [33].

Analysis of causes and consequences of pipeline failures is necessary and useful for the development of realistic risk models [34]. Risk models can be developed based on relationships between pipeline design variables and common consequences of pipeline accidents.

Another useful idea states that oil and gas pipelines can present fatal damage that leads to accidents in the form of a rupture or, more frequently, in the form of latent damage that can result in failure at a later date [35].

Applied research provides a statistical analysis approach to the frequency and consequences of gas, oil, and refined products of onshore pipelines, using data from Europe, Canada, UK, US, and Brazil [36]. In this research study, the distribution of significant failure causes is associated with pipeline parameters.

Another recent study contains a statistical analysis of accidents related to hazardous products pipeline failure; in order to identify the most common causes, the analysis comprises three classes of products among the most commonly transported through pipelines: crude oil, natural gas, and oil refined products, and highlights differences and similarities between them [37].

Among the concerns regarding the rigorous establishment of a hierarchy of the causes of incidents occurring in oil pipelines, a paper can be mentioned that proposes an expert system in onshore pipelines, highlighting failure mechanisms with the following frequency order: external corrosion, internal corrosion, third parties, erosion, material failure, and vandalism [38].

In order to avoid incidents of pipeline failure and maintain safe and reliable pipeline infrastructure, substantial research efforts have been carried out to implement pipeline leak detection and localization using different approaches [39].

There are also newer concerns that examine the relationships among environmental accidents and incidents, environmental consciousness, and financial performance [40]. In this regard, the results show that environmental consciousness has an expected significant negative effect on financial performance, whereas pipeline accidents and incidents have no expected negative effect on financial performance.

3. Materials and Methods

Based on detailed historical data regarding the incidents that occurred in Romania in the crude oil pipeline transport system, information was processed using a very developed tool of statistical methods.

In order to analyze the seasonality of the data, we graphically represented the chronogram (Figure 1) and "heat map" (Figure 2) using Excel, for the analyzed period.

When modelling this component, it is necessary to determine to what extent and in what direction the seasonality of the time series terms deviates from the central tendency, in different phases of the period, usually of the year [41] (p. 215), since in the profile literature the seasonality is investigated after the elimination of the trend [42] (p. 233).

The seasonality index represents a relative issue that expresses the intensity of the seasonal wave that characterizes the evolution of the economic process in the annual sub period j (quarter, month) [43] (p. 207). The seasonality index results, in the case of the stationary series, are generated by relating the level of sub period j or the average of the values regarding sub period j for several years to the general average return for an annual sub period, according to the formula [43] (p. 207):

$$I_j^s = \frac{\sum_{i=1}^{m} \frac{y_{ij}}{m}}{\bar{y}} \tag{1}$$

where
$$\begin{cases} i = 1, 2, \ldots m \text{ years} \\ j = 1, 2, \ldots h \text{ quarters/months} \\ \bar{y} = \text{quarterly/monthly average over the entire interval} \end{cases}$$

In the case of the trend series (non-stationary time series), it is recommended that, in a first phase, "to eliminate the trend can be achieved by relating the empirical (real) values

y_i to the (adjusted) trend values Y_i and then calculating the indices of seasonality" using the formula [44] (p. 199). Therefore, [43] (p. 207):

$$I_j^s = \frac{[\sum_i (y_{ij}/Y_{ij})] : m}{(\sum_i \sum_j y_{ij}/Y_{ij}) : mh} \qquad (2)$$

where Y_{ij} = the central trend.

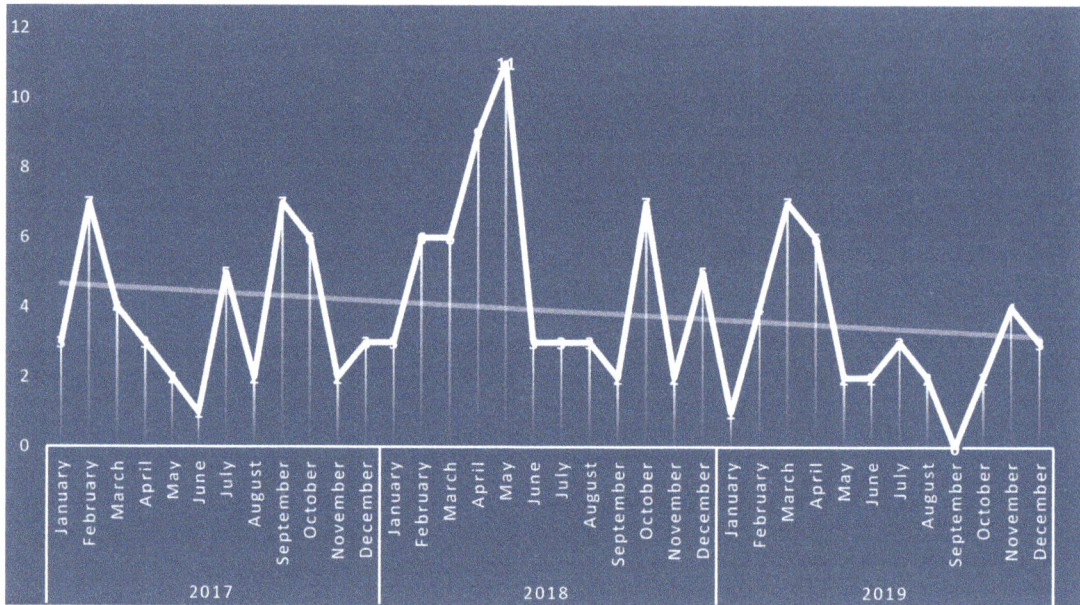

Figure 1. The number of oil-pipeline-related incidents in Romania for years 2017 to 2019 containing the chronogram and seasonality for incidents per month.

Year	January	February	March	April	May	June	July	August	September	October	November	December
2017	3	7	4	3	2	1	5	2	7	6	2	3
2018	3	6	6	9	11	3	3	3	2	7	2	5
2019	1	4	7	6	2	2	3	2	0	2	4	3

Figure 2. Heat map for incidents per month.

In the situation when $I_j^s > 1$, the evolution from "season" j is higher than the average (peak season); if $I_j^s < 1$, the evolution from "season" j is lower than the average (weak season).

To determine the seasonality indices, we used the multiplicative model. The specific stages are [41] (pp. 218–219):

(a) The ratio between the terms of the chronological series (y_{ij}) and the corresponding values of the trend (Y_{ij}), obtained by the method of moving averages or other trend analytical methods, is determined. The reports contain the seasonal component and the random component (ε_{ij}), according to the relation:

$$y_{ij}/Y_{ij} = S_j^* \cdot \varepsilon_{ij}$$

where: $i = 1, \ldots, n$; and $j = 1, \ldots, m$.

(b) The partial means ($S_j^{*'}$) are calculated on sub periods with the help of the arithmetic mean, partial means called estimators of the seasonal component:

$$S_j^{*'} = \frac{1}{n}\left[\sum \frac{y_{ij}}{Y_{ij}}\right] = \frac{1}{n}\sum(S_j^* \cdot \varepsilon_{ij}) = S_j^* \frac{\sum \varepsilon_{ij}}{n} \cong S_j^* \tag{3}$$

If the trend was not calculated based on an analytical adjustment method, the product of the estimators $S_j^{*'}$ is different from 1 ($\Pi s_j^{*'} \neq 1$), we move on to the next step.

(c) The ratios between the estimators and their average are calculated, the calculation is for each sub period (season/month); thus, the corrected estimator of the seasonal component is obtained, also called seasonality index $S_j^*(I^S_j)$ of the sub period/month (season) "j" after the relationship:

$$S^*_j = S^{*'}_j : \overline{S^{*'}_j} \tag{4}$$

meaning

$$S_j^*(I^S_j) = \left[\frac{1}{n}\sum \frac{y_{ij}}{Y_{ij}}\right] : \left(\frac{1}{m}\sum S_j^{*'}\right) = \frac{S^{*'}_j}{\frac{1}{m}(\sum S^{*'}_j)} \cdot 100 \tag{5}$$

The number of seasonality indices is equal to the number of sub periods (m).

The intensity of the seasonal wave is expressed by seasonality indices, determined according to the following formula, based on the method of reporting to the average [45] (p. 664):

$$I_j = \frac{\overline{y_j}}{\overline{y}} \cdot 100 \tag{6}$$

The interpretation of seasonality indices is similar to that of the difference (Δ_j); in other words, an index greater than or equal to 100% corresponds to a peak period, and an index less than 100% is specific to a weak period.

Moreover, the linear function was used to calculate the trend for the pipeline incidents.

The principle of linear adjustment [46] (p. 169) is based on minimizing the vertical distances between the observed (empirical) values and the theoretical (adjusted) values provided by the adjustment line, also known as the method of smaller squares, respectively, $Min = \sum(y_i - Y_i)^2$ [41] (p. 209).

The linear trend is used if it is found that the graph shows an absolutely constant upward or downward trend, verified by a small variation of the absolute changes with the moving base [44] (p. 187), [41] (p. 209).

The linear model is based on the first degree function according to the relation:

$$Y_i = a + b * t_i \tag{7}$$

where a and b are the parameters of the function that are determined from the system of normal equations, obtained by the least squares method, as follows [45] (p. 637):

$$\begin{cases} a*n + b*\sum t_i = \sum y_i \\ a*\sum t_i + b*\sum t_i^2 = \sum t_i y_i \end{cases} \tag{8}$$

and if the condition is set as $\sum t_i = 0$, the system (8) becomes:

$$\begin{cases} a*n = \sum y_i \\ b*\sum t_i^2 = \sum t_i y_i \end{cases} \tag{9}$$

hence, the parameter $a = \frac{\sum y_i}{n}$ and the parameter $b = \frac{\sum t_i y_i}{\sum t_i^2}$.

To analyze the data, descriptive statistics were used; the calculations were performed using SPSS 23.0 licensed (Statistical Package for Social Science), respectively: mean, standard deviation, minimum value, and maximum value.

To analyze whether there are differences between the mean values of each variable, the Kruskal–Wallis test was applied using SPSS 23.0 software. The Kruskal–Wallis test by ranks, Kruskal–Wallis H test (or one-way ANOVA on ranks) is a non-parametric method for testing whether samples originate from the same distribution. It is used for comparing two or more independent samples of equal or different sample sizes.

The Kruskal–Wallis test is a non-parametric test that takes into account not the absolute value of the observations but their rank, the calculation formula being the following:

$$K = 12 \times \sum \left(T_j^2 / n_j \right) / [N(N+1)] - 3(N+1) \tag{10}$$

where N = total number of observations;

T_j = total treatment modalities j.

Additionally, the calculation of the Pearson parametric correlation coefficient was taken into account. The calculation of the Pearson parametric correlation coefficient is based on the following formula:

$$R_{xy} = \frac{n \sum_{i=1}^{n} x_i y_i - (\sum_{i=1}^{n} x_i)(\sum_{i=1}^{n} y_i)}{\sqrt{\left[n \sum_{i=1}^{n} x_i^2 - (\sum_{i=1}^{n} x_i)^2 \right] \left[n \sum_{i=1}^{n} y_i^2 - (\sum_{i=1}^{n} y_i)^2 \right]}} \tag{11}$$

In order to test whether there are statistically significant differences of registered incidents depending on year of incident/incident type/product type/month/county/year of incident occurred, referring to the cause of the breakdown, the Chi-Square bivariate test was applied, the results being presented in structured tables in the Section 4. The SPSS 23.0 software was used to process data while the Chi-Square bivariate test used the following general hypothesis: H_0 = There are no statistically significant differences depending on year of incident/incident type/product type/month/county/year of incident occurred, referring to the cause of the breakdown.

In order to be able to verify this hypothesis, the following formula will be applied for the calculation of the statistics χ^2, statistics that will be calculated for a significance level of p-value $\alpha = 0.05$.

$$\chi^2 = \sum_{i=1}^{k} \left[\frac{(f_0 - f_t)^2}{f_t} \right] \tag{12}$$

where

$$\begin{cases} f_0 = observed\ frequencies \\ f_t = theoretical\ frequencies \end{cases}$$

It is analyzed if the requirements of application of the test are met, respectively:
- The sample has more than 50 statistical observations;
- There are no cells with values less than 5 or equal to zero;
- Values are absolute values and not percentages.
- The decision to reject or accept the statistical hypothesis is as follows:
- Comparing the two values (calculated with SPSS and the theoretical one, from the distribution tables); if it is observed that $\chi^2_{calculated} < \chi^2_{theoretical}$ then it results in the null hypothesis H_0 being accepted and therefore there are no statistically significant differences;
- Comparing the two values (calculated with SPSS and the theoretical one, from the distribution tables); if it is observed that $\chi^2_{calculated} > \chi^2_{theoretical}$ then it results in the null hypothesis H_0 being rejected and therefore there are statistically significant differences.

For continuous variables from the study, the Student *t* test (independent) is used to analyze the statistically significant differences, and the results are presented in the last part of the next section.

4. Results

From Figure 1 it is observed that, regarding the number of incidents per month, in the analyzed period we can say that there is certainty regarding their seasonality; respectively, the peak season is the first quarter, more specifically for 2018 and 2019, March–April. The weak seasons are represented by the summer months, predominantly. For the analyzed period, the trend of the number of incidents/month was decreasing, as can be seen in Figure 1. Although the seasonality by quarters indicates the 2nd quarter as the peak season for all years, detailed by months, atypical aspects are observed, respectively, asymmetries within a quarter. For 2017, February is the peak season, while for 2018, the peak season is represented by May and for 2019 by March. Another atypical situation is shown by the fact that, for the years 2017 and 2018, September is also the peak season, while for 2019, in September, no incidents were registered. A symmetry that must be signaled is shown by the fact that, in each of the 3 years, June represents the weak season.

We made, for the same indicator, *Number of incidents per month*, and a graph type "heat map" (Figure 2), the data being monthly in order to better see the months of the year with the highest number of such events. The figure contains values on green background (reduced number of breakdowns per month) and values on red background (increased number of incidents per month).

It can be seen that the large number of events is concentrated in spring and September–October with a maximum in 2018 in April and May.

Figure 3 shows the time series (chronogram) for the variables total cost per month and average cost per month. From Figure 3 it can be seen that, for the analyzed period, both variables had a decreasing trend. The values from Y axis refer to the local currency (1 USD = 4.2 RON).

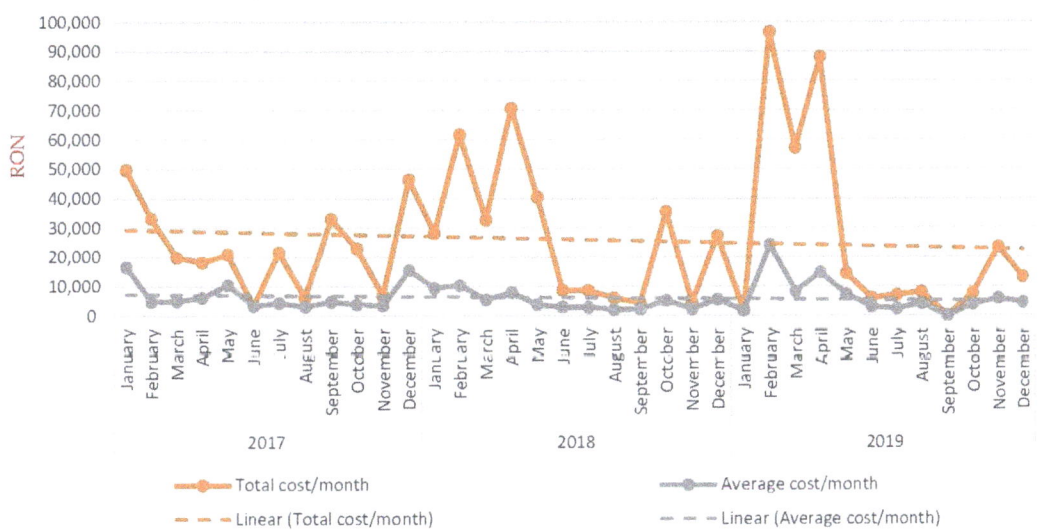

Figure 3. Chronogram, seasonality, and linear trend for total cost per month and average cost per month.

Table 2 presents descriptive statistics for the variables number of incidents, total cost/month, and average cost/month. Data are presented as: mean ± std. Deviation (minimum–maximum). To analyze whether there are differences between the mean values of each variable in column 1, the Kruskal–Wallis test was applied using SPSS software.

Table 2. Descriptive statistics and *p*-value for number of incidents, total cost/month, and average cost/month.

	2017	2018	2019	*p*-Value *
Total number of incidents/month	4 ± 2 (1–7)	5 ± 3 (2–11)	3 ± 2 (0–7)	0.182
Total cost/month (RON)	23,525.75 ± 14,843.16 (3331.05–49,662.80)	27,361.56 ± 22,425.64 (4405.09–70,687.77)	26,897.18 ± 34,105.78 (0–96,548.27)	0.710
Average cost/month (RON)	6751.52 ± 4747.47 (3114.22–16,554.27)	4942.23 ± 2877.15 (1996.36–10,278.50)	6595.06 ± 6706.59 (0–24,137.07)	0.567

* *p*-value was calculated with the Kruskal–Wallis test.

Since the *p*-values of the level are not statistically significant, based on the Kruskal–Wallis test they are over 0.05, there are no statistically significant differences between the average values of these indicators, depending on the year in which they were recorded (Table 3).

Table 3. Results for Kruskal–Wallis Test [a,b].

	Total Number of Incidents/Month	Total Cost/Month	Average Cost/Month
Chi-Square	3.405	0.686	1.135
df	2	2	2
Asymp. Sig.	0.182	0.710	0.567

[a] Kruskal–Wallis test; [b] grouping variable: year of incident registration.

Thus, the normality of the distribution of these indicators was further tested using the One-Sample Kolmogorov–Smirnov test; the *p*-value < 0.05 for all three indicators, so all of them had a normal distribution (Table 4).

Table 4. Results for normal distribution test with One-Sample Kolmogorov–Smirnov Test.

		Total Number of Incidents/Month	Total Cost/Month (RON)	Average Cost/Month (RON)
N		36	36	36
Normal Parameters [a,b]	Mean	3.92	25,928.1658	6096.2682
	Std. Deviation	2.430	24,410.74107	4950.79784
Most Extreme Differences	Absolute	0.230	0.153	0.224
	Positive	0.230	0.153	0.224
	Negative	−0.132	−0.144	−0.160
Test Statistic		0.230	0.153	0.224
Asymp. Sig. (2-tailed)		0.000 [c]	0.033 [c]	0.000 [c]

[a] Test distribution is normal; [b] calculated from data; [c] Lilliefors significance correction.

It was tested if there are statistically significant correlations between the three indicators, the results being presented in Table 5. Thus, there is a direct (positive) correlation of medium to strong intensity (0.622) that is statistically significant (*p*-value = 0.000) between the total number of incidents per month and the total cost. Moreover, Table 5 presents the results obtained with SPSS software (in fact x and y in the Formula (11) take the values of each pair in turn, for example: total number of incidents per month and total costs per month, etc.).

The following table is related to the cross-tabulation (Table 6) and summarizes the causes of incidents for each year and also this info is represented graphically in Figure 4.

Table 5. Pearson Correlation coefficients.

		Total Incidents/Month	Total Cost/Month	Average Cost/Month
Total number of incidents/month	Pearson Correlation	1	0.622 **	0.175
	Sig. (2-tailed)		0.000	0.306
	N	36	36	36
Total cost/month	Pearson Correlation		1	0.827 **
	Sig. (2-tailed)			0.000
	N		36	36
Average cost/month	Pearson Correlation			1
	Sig. (2-tailed)			
	N			36

**. Correlation is significant at the 0.01 level (2-tailed).

Table 6. The cross-tab for cause and year of incident registration.

Cause \ Year of Incidence	2017	2018	2019	Total
Corrosion	30	48	21	99
Handcrafted installation	15	9	9	33
Hole in the pipe	0	1	1	2
Metallic tap in the pipe	0	1	0	1
Accidental breakage of the pipe blower	0	0	2	2
Crack in pipe protection	0	0	1	1
Attempted pipe sectioning	0	0	2	2
Total	45	59	36	140

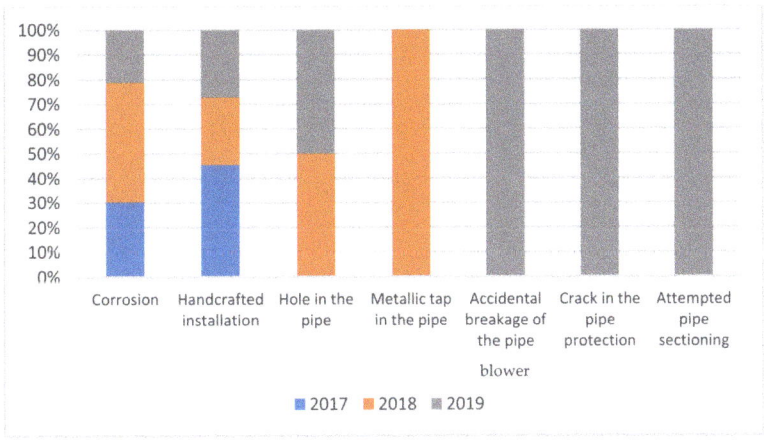

Figure 4. The annual structure of causes of incidents.

For the total analyzed period, the distribution of the incident causes is in the following table (Table 7), the most common being corrosion—70.2% and handcrafted installation—23.4% of the total causes.

Table 7. Hierarchy of incident causes for 2017–2019.

	Frequency	Percent	Valid Percent	Cumulative Percent
Corrosion	99	70.2	70.7	70.7
Handcrafted installation	33	23.4	23.6	94.3
Hole in the pipe	2	1.4	1.4	95.7
Metallic tap in the pipe	1	0.7	0.7	96.4
Accidental breakage of the pipe blower	2	1.4	1.4	97.9
Crack in pipe protection	1	0.7	0.7	98.6
Attempted pipe sectioning	2	1.4	1.4	100.0
Total	140	99.3	100.0	
Missing system	1	0.7		
Total	141	100.0		

According to the analysis of the recorded incidents, in the table below the absolute and relative frequencies of where the incidents occurred were calculated, and most of the events took place at the following pipes, marked in red in the table (Table 8).

Table 8. Recorded incidents by pipe localization and type.

	Frequency	Percent	Valid Percent	Cumulative Percent
Ø 10 3/4″ Icoana-Cartojani	4	2.8	2.8	2.8
Ø 10 3/4″ Madulari-Botorani	1	0.7	0.7	3.5
Ø 10 3/4″ Vermesti-Tg. Ocna	1	0.7	0.7	4.3
Ø 10 3/4″ F1 Bărbătești-Orleș	2	1.4	1.4	5.7
Ø 10 3/4″ F1 Barbatesti-Orlesti	1	0.7	0.7	6.4
Ø 10 3/4″ F1 Orlesti-P. Lacului	1	0.7	0.7	7.1
Ø 10 3/4″ F1 Orlesti-Poiana Lacul	1	0.7	0.7	7.8
Ø 10 3/4″ F1 P. Lacului-Siliste	1	0.7	0.7	8.5
Ø 10 3/4″ F2 Orlesti-P. Lacului	1	0.7	0.7	9.2
Ø 10 3/4″ F2 Orlesti-Poiana Lacul	2	1.4	1.4	10.6
Ø 10 3/4″ F2 Poiana Lacului-Silis	1	0.7	0.7	11.3
Ø 10 3/4″ F2 Rădinești-Orlești	1	0.7	0.7	12.1
Ø 10 3/4″ F2 Siliste-Ploiesti	3	2.1	2.1	14.2
Ø 10 3/4″ Icoana-Cartojani	2	1.4	1.4	15.6
Ø 10″ Centura	1	0.7	0.7	16.3
Ø 10″ F1 Barbatesti-Ploiesti	1	0.7	0.7	17.0
Ø 10″ Ghercesti-Icoana	1	0.7	0.7	17.7
Ø 10″ Icoana-Cartojani	1	0.7	0.7	18.4
Ø 10″ Poiana Lacului-Siliste F1	1	0.7	0.7	19.1
Ø 10″ Lascar C.-Independenta	1	0.7	0.7	19.9
Ø 12 3/4″ Cartojani-Ploiesti	3	2.1	2.1	22.0
Ø 12″ Cartojani-Ploiesti	1	0.7	0.7	22.7
Ø 12″ F1 Calareti-Ploiesti	1	0.7	0.7	23.4
Ø 14 3/4″ Brazi Refinery	1	0.7	0.7	24.1
Ø 14 3/4″ Cartojani-Ploiesti	2	1.4	1.4	25.5

Table 8. *Cont.*

	Frequency	Percent	Valid Percent	Cumulative Percent
Ø 14″ Calareti-Pitesti	2	1.4	1.4	27.0
Ø 14″ Constanta-Arpechim	2	1.4	1.4	28.4
Ø 14″ Constanta-Baraganu	1	0.7	0.7	29.1
Ø 20″ Baraganu-Calareti	2	1.4	1.4	31.2
Ø 20″ Baraganu-Călăreți	1	0.7	0.7	29.8
Ø 20″ Călăreți-Pitești	1	0.7	0.7	31.9
Ø 20″ Constanta-Baraganu	16	11.3	11.3	43.3
Ø 24″ Baraganu-Calareti	15	10.6	10.6	53.9
Ø 24″ Calareti-Pietrosani	2	1.4	1.4	55.3
Ø 24″ Calareti-Ploiesti	7	5.0	5.0	60.3
Ø 24″ Constanta-Petromidia	3	2.1	2.1	62.4
Ø 28″ Constanta-Baraganu	5	3.5	3.5	66.0
Ø 4 + 5″ Surani-Matita	1	0.7	0.7	66.7
Ø 4″ Recea-Mislea	1	0.7	0.7	67.4
Ø 4″+ 5″ Surani-Magurele	2	1.4	1.4	68.8
Ø 4″+ 5″ Surani-Matita	2	1.4	1.4	70.2
Ø 5 9/16″ G. Vitioarei-Ploiesti	2	1.4	1.4	71.6
Ø 5 9/16″ Gura Vit.-Magurele	1	0.7	0.7	72.3
Ø 5 9/16″ Păcureți-Măgurele	1	0.7	0.7	73.0
Ø 5″ GuraVitioarei-Boldești	1	0.7	0.7	73.8
Ø 6 + 5″ GuraVitioarei-Boldeșt	1	0.7	0.7	74.5
Ø 6 5/8″ Baicoi-Banesti	1	0.7	0.7	75.2
Ø 6 5/8″ F1 Ticleni-Ploiesti	5	3.5	3.5	78.7
Ø 6 5/8″ Grindu-Urziceni	2	1.4	1.4	80.1
Ø 6 5/8″ Izvoru-Izvoruracordare	2	1.4	1.4	81.6
Ø 6 5/8″ Moreni-Ploiesti	1	0.7	0.7	82.3
Ø 6 5/8″ Ochiuri-Moreni	1	0.7	0.7	83.0
Ø 6 5/8″ Padure II-Buda	2	1.4	1.4	84.4
Ø 6 5/8″ Teis-Moreni	3	2.1	2.1	86.5
Ø 6 5/8″ Urlati-Ploiesti	3	2.1	2.1	88.7
Ø 6 5/8″ Urziceni-Albesti	6	4.3	4.3	92.9
Ø 6 5/8″ Urziceni-Ploiesti	2	1.4	1.4	94.3
Ø 6 5/8″ Lact-Icoana	1	0.7	0.7	95.0
Ø 6″ Izvoru-Izvoruracordare	1	0.7	0.7	95.7
Ø 8″ Lucacesti-Vermesti	1	0.7	0.7	96.5
Ø 8 + 6″Oprisenesti-Ciresu	1	0.7	0.7	97.2
Ø 8 5/8″ Boldesti-Raf Teleajen	1	0.7	0.7	97.9
Ø 8 5/8″ Centura	1	0.7	0.7	98.6
Ø 8″ Lucacesti-Vermesti	1	0.7	0.7	100.0
Ø 8″ Lucăcești-Vermești	1	0.7	0.7	99.3
Total	141	100.0	100.0	

The table below includes the counties with the highest number of pipe incidents in the analyzed period, marked in red in the table (Table 9).

Table 9. Recorded incidents by county.

	Frequency	Percent	Valid Percent	Cumulative Percent
Ialomita	10	7.1	7.1	7.1
Valcea	6	4.3	4.3	11.3
Arges	15	10.6	10.6	22.0
Calarasi	19	13.5	13.5	35.5
Prahova	32	22.7	22.7	58.2
Ilfov	4	2.8	2.8	61.0
Dambovita	17	12.1	12.1	73.0
Constanta	23	16.3	16.3	89.4
Bacau	4	2.8	2.8	92.2
Teleorman	2	1.4	1.4	93.6
Gorj	3	2.1	2.1	95.7
Giurgiu	3	2.1	2.1	97.9
Galati	1	0.7	0.7	98.6
Olt	1	0.7	0.7	99.3
Braila	1	0.7	0.7	100.0
Total	141	100.0	100.0	

Among the transported oil products, the product most affected by the incidents was domestic crude oil (55.3%), followed by imported crude oil product with 41.1% (Table 10).

Table 10. Recorded incidents by transported product.

	Frequency	Percent	Valid Percent	Cumulative Percent
Crude oil (from Romania)	78	55.3	55.3	55.3
Imported crude oil	58	41.1	41.1	96.5
Gasoline	5	3.5	3.5	100.0
Total	141	100.0	100.0	

The most common incident is the technical one with 73% of the total, the difference being represented by the intentionally caused incident, with 27% of the total (Table 11).

Table 11. Recorded incidents by incident type.

	Frequency	Percent	Valid Percent	Cumulative Percent
Technical	103	73.0	73.0	73.0
Intentionally caused	38	27.0	27.0	100.0
Total	141	100.0	100.0	

It is worth highlighting the high frequency of "intentionally caused" incidents. The fact that petroleum products (representing important and expensive conventional resources) are transported on these pipelines, which, through excise duty, are sold at significantly higher price values, explains the temptation to use artisanal installations through which to divert substantial quantities. This situation reveals the continuous concern of the decision makers in proposing ample measures and actions for monitoring on the ground and in the air, through which to prevent such incidents/provoked breakdowns.

In order to perform the Chi-Square test, cross-tabulation is used again. Cross-tabulation greatly helps in research by identifying patterns, trends, and the correlation between parameters. Therefore, a cross-tabulation is made regarding the year of incident registration and causes of incident for each analyzed year. Table 12 was built based on it.

Table 12. The cross-tab for year of incident registration and cause of incident.

Year of Incident Registration	Cause	Corrosion	Handcrafted Installation	Hole in the Pipe	Metallic Tap in the Pipe	Accidental Breakage of the Pipe blower	Crack in Pipe Protection	Attempted Pipe Sectioning	Total
2017		30	15	0	0	0	0	0	45
2018		48	9	1	1	0	0	0	59
2019		21	9	1	0	2	1	2	36
Total		99	33	2	1	2	1	2	140

The defining results of the Chi-Square test can be found in Table 13.

Table 13. Results for Chi-Square bivariate test.

	Value	df	Asymptotic Sig. (2-Sided)
Pearson Chi-Square	22.360	12	0.034
Likelihood Ratio	22.440	12	0.033
Linear-by-Linear Association	7.363	1	0.007
N of Valid Cases	140		

From Table 13, because the p-value is <0.05, the null hypothesis H_0 is rejected and therefore there are significant differences depending on the cause of the incident related to the year in which the incident occurred (one of the observable differences in Table 12—cross-tab being the much higher number of incidents in 2016 caused by handcrafted installations).

Another cross-tabulation concerns the incident type–incident cause pair. The cross-tabulation results are mentioned in Table 14.

Table 14. The crosstab for incident type and cause.

Incident Type	Cause	Corrosion	Handcrafted Installation	Hole in the Pipe	Metallic Tap in the Pipe	Accidental Breakage of the Pipe blower	Crack in Pipe Protection	Attempted Pipe Sectioning	Total
Technical		99	0	0	0	2	1	0	102
Intentionally caused		0	33	2	1	0	0	2	38
Total		99	33	2	1	2	1	2	140

Table 15 contains the Chi-Square test, and in this case it is observed that the null hypothesis H_0 is rejected (p-value = 0.000) and therefore there are statistically significant differences depending on the cause of the pipe incident and the incident type for the analyzed period.

Another pair of elements refers to the type of product transported through the pipeline and the incident cause. The cross-tabulation results are included in Table 16.

Table 15. The results for Chi-Square bivariate test.

	Value	df	Asymptotic Sig. (2-Sided)
Pearson Chi-Square	140.000	6	0.000
Likelihood Ratio	163.709	6	0.000
Linear-by-Linear Association	39.822	1	0.000
N of Valid Cases	140		

Table 16. The cross-tab for product type and cause.

Product Type	Cause	Corrosion	Handcrafted Installation	Hole in the Pipe	Metallic Tap in the Pipe	Accidental Breakage of the Pipe blower	Crack in Pipe Protection	Attempted Pipe Sectioning	Total
Crude oil (from Romania)		53	20	0	0	2	1	2	78
Imported crude oil		41	13	2	1	0	0	0	57
Gasoline		5	0	0	0	0	0	0	5
Total		99	33	2	1	2	1	2	140

Thus, Table 17 describes the corresponding Chi-Square test.

Table 17. The results for Chi-Square bivariate test.

	Value	df	Asymptotic Sig. (2-Sided)
Pearson Chi-Square	10.380	12	0.583
Likelihood Ratio	14.384	12	0.277
Linear-by-Linear Association	2.564	1	0.109
N of Valid Cases	140		

The conclusion is that regarding the cause of the pipe incident and the type of product, the results indicate that there are no statistically significant differences.

Moreover, there are no differences depending on the cause and the month of the year when the incident occurred, according to the results of the Chi-Square test in the table below (Table 18).

Table 18. The results for Chi-Square bivariate test.

	Value	df	Asymptotic Sig. (2-Sided)
Pearson Chi-Square	65.515	66	0.494
Likelihood Ratio	49.988	66	0.929
Linear-by-Linear Association	0.866	1	0.352
N of Valid Cases	140		

In addition, there are no differences depending on the cause and the county in which the incident occurred, according to the results of the Chi-Square test in the table below (Table 19).

Last but not least, there are no differences depending on the cause and the year in which the incident occurred, according to the results of the Chi-Square test in the table below (Table 20).

Table 19. The results for Chi-Square bivariate test.

	Value	df	Asymptotic Sig. (2-Sided)
Pearson Chi-Square	89.482	84	0.321
Likelihood Ratio	62.467	84	0.962
Linear-by-Linear Association	0.309	1	0.579
N of Valid Cases	140		

Table 20. The results for Chi-Square bivariate test.

	Value	df	Asymptotic Sig. (2-Sided)
Pearson Chi-Square	840.000	834	0.435
Likelihood Ratio	234.739	834	1.000
Linear-by-Linear Association	2.497	1	0.114
N of Valid Cases	140		

To analyze whether there are statistically significant differences depending on the type of incident between the average values of the other indicators in the study, the Student's t test was applied, the results being presented in the following tables (Tables 21 and 22).

Table 21. Descriptive statistics of Student's t test.

	Incident Type	N	Mean	Std. Deviation	Std. Error Mean
Cause	Technical	102	1.13	0.740	0.073
	Intentionally caused	38	2.37	1.172	0.190
Cost	Technical	103	7344.4314	11,934.05523	1175.89740
	Intentionally caused	38	4656.2511	3531.61935	572.90431
Incident data	Technical	103	30 May 2018	276.07:22:36.543	27.05:24:32.304
	Intentionally caused	38	15 May 2018	326.05:09:54.642	52.22:03:29.885
County	Technical	103	5.42	2.721	0.268
	Intentionally caused	38	6.29	3.075	0.499
Month	Technical	103	6.17	3.552	0.350
	Intentionally caused	38	5.87	3.086	0.501
Transported product	Technical	103	1.50	0.592	0.058
	Intentionally caused	38	1.42	0.500	0.081

Table 22. Results for Student's *t* test.

		Levene's Test for Equality of Variances		*t*-Test for Equality of Means					95% Confidence Interval of the Difference	
		F	Sig.	t	df	Sig. (2-Tailed)	Mean Difference	Std. Error Difference	Lower	Upper
Cause	Equal variances assumed	6.975	0.009	−7.444	138	0.000	−1.241	0.167	−1.571	−0.911
	Equal variances not assumed			−6.089	48.419	0.000	−1.241	0.204	−1.651	−0.831
Cost	Equal variances assumed	4.413	0.037	1.364	139	0.175	2688.18	1970.92749	−1208.69361	6585.05423
	Equal variances not assumed			2.055	135.174	0.042	2688.18	1308.03442	101.32089	5275.03973
Incident type	Equal variances assumed	5.628	0.019	0.271	139	0.787	14:22:52:18.27	55.02:58:48.902	−94.00:53:43.219	123.22:38:19.765
	Equal variances not assumed			0.251	57.712	0.803	14:22:52:18.27	59.12:16:57.600	−104.04:27:02.096	134.02:11:38.642
County	Equal variances assumed	0.462	0.498	−1.630	139	0.105	−0.872	0.535	−1.930	0.186
	Equal variances not assumed			−1.540	59.657	0.129	−0.872	0.566	−2.005	0.261
Month	Equal variances assumed	1.964	0.163	0.470	139	0.639	0.306	0.652	−0.982	1.595
	Equal variances not assumed			0.502	75.469	0.617	0.306	0.611	−0.910	1.523
Transported product	Equal variances assumed	2.949	0.088	0.776	139	0.439	0.084	0.108	−0.130	0.297
	Equal variances not assumed			0.838	77.581	0.404	0.084	0.100	−0.115	0.283

The above results show that, if we group the study data according to the type of incident, there are statistically significant differences between the averages of the following variables in the study: cause of incident, total cost of incident, and incident type (p-value < 0.05), and for the transported product, a level of statistical significance of 91.2%. Table 23 contains the matrix of Pearson parametric correlation coefficients.

Table 23. Pearson correlation coefficients.

		Cause	Cost	County	Month	Product	Incident Type
Cause	Pearson Correlation	1	0.134	0.047	−0.079	−0.136	**0.535 ****
	Sig. (2-tailed)		0.114	0.580	0.354	0.110	**0.000**
	N	140	140	140	140	140	140
Cost	Pearson Correlation		1	−0.075	**−0.172 ***	−0.006	−0.115
	Sig. (2-tailed)			0.380	**0.041**	0.945	0.175
	N		141	141	141	141	141
County	Pearson Correlation			1	0.110	0.069	0.137
	Sig. (2-tailed)				0.194	0.414	0.105
	N			141	141	141	141
Month	Pearson Correlation				1	0.105	−0.040
	Sig. (2-tailed)					0.213	0.639
	N				141	141	141
Product	Pearson Correlation					1	−0.066
	Sig. (2-tailed)						0.439
	N					141	141
Incident type	Pearson Correlation						1
	Sig. (2-tailed)						
	N						141

** Correlation is significant at the 0.01 level (2-tailed). * Correlation is significant at the 0.05 level (2-tailed).

The following statistically significant correlations are the result:
- There is a statistically significant direct correlation of average intensity (Pearson correlation coefficient = 0.535) between the type of incident and its cause;
- There is a statistically significant inverse correlation of low intensity (Pearson correlation coefficient = −0.172) between the month of incident and its cost.

5. Conclusions

This paper presents a statistical analysis of the main oil pipeline system from Romania in terms of failure event rates and the hierarchy of the main causes of incidents.

The causes identified and analyzed were classified into seven categories: corrosion; handcrafted (artisanal) installation; hole in the pipe; metallic tap in the pipe; accidental breakage of the pipe blower; crack in pipe's protection; attempted pipe sectioning.

Major pipeline incident events often result in injuries, fatalities, property damage, fires, explosions, and release of hazardous materials. Because of these multiple consequences, detailed statistical analyses are needed related to the causes that generated these events.

In this sense, any analysis has to start from the following description regarding the general condition of the oil and gas transport systems: it is known that most European pipeline systems were built in the 1960s and 1970s, while in 2019 less than 2% of the pipelines were 10 years old or less and 70% were over 40 years old, and 40% of the pipeline networks worldwide have reached their projected 20-year service lifetime. This situation is similar in North America, Russia, and even Australia.

Over time, general and specific studies have been conducted on the analysis of incidents in oil and gas pipelines around the world. The most important studies were conducted in the United States (via PHMSA) and in Europe (by UKOPA and EGIG). The present study introduces representative elements in the case of incidents occurring in the national transport system of petroleum products in Romania in order to initiate useful steps in harmonizing the causes of these incidents with the analysis and recommendations of international professional associations in this field.

The main ideas and findings of the present analysis can be presented as follows:

- The most common causes refer to corrosion (especially internal corrosion) and hand-crafted (artisanal) installations (in this last case, the decision makers are inclined to intervene promptly by promoting ground and air patrol missions);
- There is a linear tendency to reduce incidents due to artisanal installations (starting with 2017 when monitoring was started by patrolling crews with people, for land security and day by day checks);
- Most incidents occurred in pipes with large diameters and while transporting imported crude oil;
- The counties most affected by the incidents are represented by points of major interest (Constanta, with the crude oil terminal that ensures the supply of refineries with imported crude oil transported by vessels; Prahova, through the Brazi refinery that processes both crude and imported crude oil; Calarasi and Ialomita, as nodal points that have pumping stations for the transport of crude oil to Moldova and Muntenia);
- There is a seasonality that shows that the 2nd quarter of each year (especially the months of March and April) presents an increased number of events; the explanation is given by the fact that these months are marked by numerous days with precipitation, and the patrol missions are hampered by the climatic conditions so that the incidents that have as a source the artisanal installations are more numerous; we meet the same situation in September.
- There is a seasonality that shows up as well in the 2nd quarter of 2018 and 2019 (especially the months of February and April) for *total cost per month* and *average cost per month* due to the highest number of incidents in April and May 2018 (9, respectively, 11) from the entire analyzed period (according to the heat map from Figure 2);
- Based on the results of the Kruskal–Wallis test, there are no differences between the studied years depending on number of incidents, total cost/month, and average cost/month;
- There is positive statistical significance and medium through strong correlation between *total cost/month* and *total number of incidents/month*;
- There is positive statistical significance and strong correlation between *average cost/month* and *total cost/month*;
- According to the results of the Chi-Square bivariate test:
    • There are statistically significant differences between the years from the study depending on the cause of incident and incident type;
    • There are no statistically significant differences between the years from the study depending on product type, month in which incident occurred, county in which the incident occurred, cause of incident.
- According to the results of Student's *t* test, there are statistically significant differences depending on *incident type* between: *the means of incident causes, the means of total cost,* and *the means of* incident data;
- There is positive statistical significance and medium correlation between *incident type* and *incident cause*;
- There is negative statistical significance and weak correlation between the *month of the incident* and *the total cost.*

Taking into account these observations and the fact that crude oil is considered to be the "black" gold for a country's economy, some measures have already been initiated, to which the authors add proposals to generate a more efficient use of crude oil resources (especially) using pipeline network. This more efficient use takes into account, on the one hand, natural economic requirements (cost of interventions, costs of replacing equipment and pipelines affected by accidents, monitoring costs, and costs of reducing or eliminating adverse effects on soil and water), and on the other hand, the creation of a much safer technical infrastructure to ensure the protection of the environment.

The concept targeted in the paper, including through statistical analysis, but also from the need to implement safe practices in order to prevent, detect, and mitigate incidents that may occur in the case of pipelines, refers to pipeline integrity. This approach chronologically includes the stages of prevention, detection, and mitigation. Each of these steps can significantly help reduce the negative financial, social, and environmental effects that any incident in this sector can generate at any given time.

Specifically, prevention involves: avoiding geo-hazards along pipelines; adequately protecting pipelines against corrosion; monitoring operating pressures; inspection of pipelines; and properly training all the operators and workers involved in the process.

Practically, detection deals with: external detection systems comprising sensors, imaging (with cameras, using drones and maybe helicopters), and patrols (with cars, helicopters, and drones); internal detection systems that check the commodity pressure and/or flow in the pipes, statistical analyses regarding the condition of the pipes made automatically by specialized interfaces. Mitigation aims to *locate* the area where there are spills, *recover*, by which quick measures are taken (maximum 6–8 h) to eliminate the effects generated by incidents, and, respectively, *clean up*, which refers to the cleaning of the place where commodity leaks have occurred.

Therefore, on the one hand, it is necessary to consider the continuation and development of the modernization programs initiated as follows: upgrading the hardware and software of the existing SCADA system (type MicroSCADA 8.4.3, produced and installed by ABB ENERGY INFORMATION SYSTEMS GMBH Germany, consisting of five Base System 1 and 2 servers-redundant, Frontend 1- and 2-redundant, and a remote access server); modernization of the cathodic protection system of the pipelines (currently consisting of a number of 218 cathodic protection stations—not integrated in a unitary automated system—located on the route of the main and local pipelines, respectively); implementation, for the first time, of a leak detection and location system (leak detection type); intelligent excavation of the pipeline system, respectively, through reconstruction by guided drilling (horizontal), of some important route segments, from the category of works of art (as is the case of crossing watercourses, as the currently established solution over crossing perpetually raises issues of the order of securing the supporting elements).

On the other hand, it is necessary to carry out concrete measures for strategic lines of action:

1. Improvements of the national transport system by the implementation of the leak detection and location system, modernization of the cathodic protection system and supervisory, control, and data acquisition system (by developing existing SCADA system), and renewal of the pipeline network based on field data monitoring;
2. Economic efficiency improvements by reduction of technological consumption within the storage and transport processes, minimization of energy, fuel, and lubricant consumptions, and reduction of the operating costs;
3. Interconnection of the national crude oil pipeline transport system to the *Regional and European Systems* based on the implementation of Constanța–Pitești–Pancevo *Project*—an alternative crude oil transport solution in order to supply the Pancevo refinery (Serbia). This project has the following features:
   - Total length of the pipe—760 km;
   - Transport capacity—7.5 million tons/year;
   - Only the section pipe Pitești–Naidaș–Pancevo (440 km) needs to be built; the section pipe Constanta–Pitesti (320 km) is already built. Constanta is the main oil supply hub (for imported crude oil) in Eastern Europe and the Balkan countries.

4. To comply with legal requirements applicable to the organization and to ensure a working environment in safe conditions, organizational, administrative, and financial efforts have to be continued to recertify the management systems already functional in the company (ISO 9001:2015; ISO 14001: 2015; ISO 45001:2018; ISO 50001:2018). Based on these recertified systems (last time in 2019), the inclusion of new procedures and the updating of existing ones regarding the inspection and maintenance plans of the pipeline systems must be considered.

Author Contributions: Conceptualization, C.P. and M.R.G.; Methodology, C.P. and M.R.G.; Validation, C.P. and M.R.G.; Formal Analysis, C.P. and M.R.G.; Investigation, C.P. and M.R.G.; Writing—Original Draft Preparation, C.P. and M.R.G.; Writing—Review & Editing, C.P. and M.R.G. All authors have read and agreed to the published version of the manuscript.

Funding: This research received no external funding.

Conflicts of Interest: The authors declare no conflict of interest.

References

1. Ramírez-Camacho, J.; Carbone, F.; Pastor, E.; Bubbico, R.; Casal, J. Assessing the consequences of pipeline accidents to support land-use planning. *Saf. Sci.* **2017**, *97*, 34–42. [CrossRef]
2. Khan, F.; Yarveisy, R.; Abbassi, R. Risk-based pipeline integrity management: A road map for the resilient pipelines. *Pipeline Sci. Eng.* **2021**, *1*, 74–87. [CrossRef]
3. Miao, C.; Zhao, J. Risk analysis for the urban buried gas pipeline with fuzzy comprehensive assessment method. *J. Press. Vessel Technol.* **2012**, *134*. [CrossRef]
4. Green, K.P.; Jackson, T. Pipelines Are the Safest Way to Transport Oil and Gas. Fraser Research Bulletin by the Faser Institute, Canada. 2015. Available online: https://www.fraserinstitute.org/article/pipelines-are-safest-way-transport-oil-and-gas (accessed on 3 May 2021).
5. Grigoriev, A.; Grigorieva, N. The valve location problem: Minimizing environmental damage of a spill in long oil pipelines. *Comput. Ind. Eng.* **2009**, *57*, 976–982. [CrossRef]
6. Belvederesi, C.; Thompson, M.; Komers, P. Statistical analysis of environmental consequences of hazardous liquid pipeline accidents. *Heliyon* **2018**, *4*, e00901. [CrossRef]
7. Green, K.P.; Jackson, T. Safety in the Transportation of Oil and Gas: Pipelines or Rail? *Fraser Research Bulletin by the Faser Institute, Canada*. 2015. Available online: https://www.fraserinstitute.org/research/safety-transportation-oil-and-gas-pipelines-or-rail (accessed on 8 May 2021).
8. Yeung, P.; White, B.; Ziccardi, M.; Chilvers, B.L. What Helps Oiled Wildlife Responders Care for Animals While Minimizing Stress and Compassion Fatigue. *Animals* **2021**, *11*, 1952. [CrossRef] [PubMed]
9. Anderson, D.A. Natural Gas Transmission Pipelines: Risks and Remedies for Host Communities. *Energies* **2020**, *13*, 1873. [CrossRef]
10. Feng, Q. Pipeline Failure Cause Theory: A New Accident Characteristics, Quantification, and Cause Theory. In *Failure Analysis*; Huang, Z.-M., Hemeda, S., Eds.; IntechOpen: London, UK, 2018. [CrossRef]
11. Furchtgott-Roth, D. *Pipelines Are Safest for Transportation of Oil and GAS*; Manhattan Institute for Policy Research: New York, NY, USA, 2013.
12. Lam, C. *Statistical Analysis of Historical Pipeline Incident Data with Application to the Risk Assessment of Onshore Natural Gas Transmission Pipelines*; The University of Western Ontario: London, ON, Canada, 2015.
13. Lam, C.; Zhou, W. Statistical analyses of incidents on onshore gas transmission pipelines based on PHMSA. *Int. J. Press. Vessel Pip.* **2016**, *145*, 29–40. [CrossRef]
14. Cech, M.; Davis, P.; Guijt, W.; Haskamp, A.; Huidobro Barrio, I. Performance of European cross-country oil pipelines. In *Statistical Summary of Reported Spillages in 2019 and Since 1971*; Report CONCAWE: Brussels, Belgium, 2021; No. 4.
15. Azevedo, C.R.F. Failure Analysis of the crude oil pipeline. *Eng. Fail. Anal.* **2007**, *14*, 978–994. [CrossRef]
16. Kheraj, S. A History of Oil Spills on Long-Distance Pipelines in Canada. *Can. Hist. Rev.* **2020**, *101*, 161–191. [CrossRef]
17. Chen, X.; Feng, Q.; Shui, B.; Li, B.; Hao, J.; Teng, T. The development and challenges of pipeline failure analysis in China. In Proceedings of the 7th International Pipeline Conference (IPC), Calgary, AB, Canada, 29 September–3 October 2008. [CrossRef]
18. Pipeline and Hazardous Materials Safety Administration. History of PHMSA Incident Reporting Criteria. 2017. Available online: http://www.phmsa.dot.gov/pipeline/library/data-stats (accessed on 20 June 2021).
19. National Transportation Safety Board. Pipeline Accident Reports. 2021. Available online: https://www.ntsb.gov/investigations/accidentreports/pages/pipeline.aspx (accessed on 20 June 2021).
20. Rezaei, H.; Ryan, B.; Stoianov, I. Pipe Failure Analysis and Impact of Dynamic Hydraulic Conditions in Water Supply Networks. *Procedia Eng.* **2015**, *119*, 253–262. [CrossRef]

21. Bubbico, R.; Carbone, F.; Ramirez-Camacho, J.G.; Pastor, E.; Casal, J. Conditional probabilities of post-release events for hazardous materials pipelines. *Process Saf. Environ. Prot.* **2016**, *104*, 95–110. [CrossRef]
22. Vtorushina, A.N.; Anishchenko, Y.V.; Nikonova, E.D. Risk Assessment of Oil Pipeline Accidents in Special Climatic Conditions. *IOP Conf. Ser. Earth Environ. Sci.* **2017**, *66*, 012006. [CrossRef]
23. Perez-Suarez, A.F.; Cabrales, S.; Amaya-Gomez, R.; Munoz, F. Model for optimal sectioning of hydrocarbon transportation pipelines by minimization of the expected economic losses. *J. Loss Prev. Process Ind.* **2019**, *62*, 103939. [CrossRef]
24. Medina, H.; Arnaldos, J.; Casal, J.; Bonvicini, S.; Cozzani, V. Risk-based optimization of the design of onshore pipeline shutdown systems. *J. Loss Prev. Process Ind.* **2012**, *25*, 489–493. [CrossRef]
25. Restrepo, C.; Simonoff, J.S.; Zimmerman, R. Causes, cost consequences and risk implications of accidents in US hazardous liquid pipeline infrastructure. *Int. J. Crit. Infrastruct. Prot.* **2009**, *2*, 38–50. [CrossRef]
26. Shin, S.; Lee, G.; Ahmed, U.; Lee, Y.; Na, J.; Han, C. Risk-based underground pipeline safety management considering corrosion effect. *J. Hazard. Mater.* **2018**, *342*, 279–289. [CrossRef]
27. Zhang, P.; Chen, X.; Fan, C. Research on a Safety Assessment Method for Leakage in a Heavy Oil Gathering Pipeline. *Energies* **2020**, *13*, 1340. [CrossRef]
28. Shahriar, A.; Sadiq, R.; Tesfamariam, S. Risk analysis for oil & gas pipelines: A sustainability assessment approach using fuzzy based bow-tie analysis. *J. Loss Prev. Process Ind.* **2012**, *25*, 505–523.
29. Bonvicini, S.; Antonioni, G.; Morra, P.; Cozzani, V. Quantitative assessment of environmental risk due to accidental spills from onshore pipelines. *Process Saf. Environ. Prot.* **2015**, *93*, 31–49. [CrossRef]
30. Khaled, A.; Alrushaid, S.; Almansouri, J.; Alrashed, A. Using fault tree analysis in the Al-Ahmadi town gas leak incidents. *Saf. Sci.* **2015**, *79*, 184–192. [CrossRef]
31. de Rujiter, A.; Guldenmund, F. The bowtie method: A review. *Saf. Sci.* **2016**, *88*, 211–218. [CrossRef]
32. Lu, L.L.; Liang, W.; Zhang, L.B.; Zhang, H.; Lu, Z.; Shan, J.Z. A comprehensive risk evaluation method for natural gas pipelines by combining a risk matrix with a bow-tie model. *J. Nat. Gas Sci. Eng.* **2015**, *25*, 124–133. [CrossRef]
33. Edjossan-Sossou, A.M.; Deck, O.; Al Heib, M.; Verdel, T. A decision-support methodology for assessing the sustainability of natural risk management strategies in urban areas. *Nat. Hazards Earth Syst. Sci.* **2014**, *14*, 3207–3230. [CrossRef]
34. Belvederesi, C.; Dann, M.R. Statistical analysis of failure consequences for oil and gas pipelines. *Int. J. Saf. Secur. Eng.* **2017**, *7*, 103–112. [CrossRef]
35. Worthington, W. Monitoring for transient pressures in pipelines. In *Pipelines 2005: Optimizing Pipeline Design, Operations, and Maintenance in Today's Economy*; Pipeline Division Specialty Conference: Houston, TX, USA, 2005; pp. 886–898. [CrossRef]
36. Cunha, S.B. A review of quatitative risk assessment of onshore pipelines. *J. Loss Prev. Process Ind.* **2016**, *44*, 282–298. [CrossRef]
37. Bubbico, R. A statistical analysis of causes and consequences of the release of hazardous materials from pipelines. The influence of layout. *J. Loss Prev. Process Ind.* **2018**, *56*, 458–466. [CrossRef]
38. Castellanos, V.; Albiter, A.; Hernández, P.; Barrera, G. Failure analysis expert system for onshore pipelines. Part—I: Structured database and knowledge acquisition. *Expert Syst. Appl.* **2011**, *38*, 11085–11090. [CrossRef]
39. Adegboye, M.A.; Fung, W.-K.; Karnik, A. Recent Advances in Pipeline Monitoring and Oil Leakage Detection Technologies: Principles and Approaches. *Sensors* **2019**, *19*, 2548. [CrossRef] [PubMed]
40. Denommee-Gravel, V.; Kim, K. Pipeline Accidents and Incidents, Environmental Consciousness, and Financial Performance in the Canadian Energy Industry. *Sustainability* **2019**, *11*, 3275. [CrossRef]
41. Baron, T.; Biji, E.; Tövisi, L.; Wagner, P.; Isaic-Maniu, A.; Korka, M.; Porojan, D. *Statistică Teoretică Și Economică*; Didactică și Pedagogică: București, Romania, 1996.
42. Tașnadi, A. *Econometrie-Ediție Revizuită Și Adăugită*; ASE: București, Romania, 2005.
43. Isaic-Maniu, A.; Pecican, E.; Ștefănescu, D.; Vodă, V.G.; Wagner, P. *Dicționar de Statistică Generală*; Economică: București, Romania, 2003.
44. Baron, T.; Anghelache, C.; Țițan, E. *Statistică*; Economică: București, Romania, 1996.
45. Andrei, T. *Statistică și Econometrie*; Economică: București, Romania, 2003.
46. Langlois, G.; Granier, C.; Bauval, M.; Guilbaud, B. *Analyse Statistique Probabilites*; Foucher: Paris, France, 1985.

Article

The Role of Globalization, Economic Growth and Natural Resources on the Ecological Footprint in Thailand: Evidence from Nonlinear Causal Estimations

Hafezali Iqbal Hussain [1,2], Muhammad Haseeb [1], Fakarudin Kamarudin [3,4,*], Zdzisława Dacko-Pikiewicz [5] and Katarzyna Szczepańska-Woszczyna [5]

[1] Taylor's Business School, Taylor's University Lakeside Campus, 1 Jalan Taylors, Subang Jaya 47500, Malaysia; hafezali.iqbalhussain@taylors.edu.my (H.I.H.); muhammad.haseeb@taylors.edu.my (M.H.)
[2] University of Economics and Human Sciences in Warsaw, Okopowa 59, 01-043 Warsaw, Poland
[3] School of Business and Economics, Universiti Putra Malaysia, Serdang 43400, Malaysia
[4] EIS-UPMCS Centre for Future Labour Market Studies, SOCSO, Putrajaya 62100, Malaysia
[5] Department of Management, Faculty of Applied Sciences, WSB University, 41-300 Dabrowa Górnicza, Poland; zdacko@wsb.edu.pl (Z.D.-P.); kszczepanska@wsb.edu.pl (K.S.-W.)
* Correspondence: fakarudin@upm.edu.my

Citation: Hussain, H.I.; Haseeb, M.; Kamarudin, F.; Dacko-Pikiewicz, Z.; Szczepańska-Woszczyna, K. The Role of Globalization, Economic Growth and Natural Resources on the Ecological Footprint in Thailand: Evidence from Nonlinear Causal Estimations. *Processes* 2021, *9*, 1103. https://doi.org/10.3390/pr9071103

Academic Editor: Lucian-Ionel Cioca

Received: 11 May 2021
Accepted: 21 June 2021
Published: 25 June 2021

Publisher's Note: MDPI stays neutral with regard to jurisdictional claims in published maps and institutional affiliations.

Copyright: © 2021 by the authors. Licensee MDPI, Basel, Switzerland. This article is an open access article distributed under the terms and conditions of the Creative Commons Attribution (CC BY) license (https://creativecommons.org/licenses/by/4.0/).

Abstract: The environmental issue has become a global problem that needs to be examined frequently, motivating researchers to investigate it. Thus, the present study has investigated the asymmetric impact of globalization, economic growth and natural resources on the ecological footprint in the presence of environmental Kuznets curve (EKC) in Thailand. The study has used annual time series data from 1970 to 2018. The study applied a novel method of nonlinear autoregressive distributive lag (ARDL). In particular, the current study has investigated the effect of positive and negative shocks on the independent variable on the dependent variable. The findings have confirmed that the effect of globalization and natural resources are significant and nonlinear. However, the effect of negative shocks of globalization and natural resources is more dominant on the ecological footprint in Thailand than the positive shocks of both variables. Moreover, the present study has also tested the presence of EKC in Thailand, and the findings confirm the presence of an inverted U-shape curve in the Thailand economy.

Keywords: globalization; economic growth; natural resources; ecological footprint; environmental Kuznets curve; Thailand

1. Introduction

The degradation of the environment is one of the most urgent challenges facing the global community. Resource utilization at higher rates could impact the environment. With this backdrop of decreasing resources, climate change is seen as one of the major challenges of the modern human race. It is safe to mention that everyone is responsible for this progressive worsening of living conditions regardless of any division of developed or developing countries. It is a widely known fact that natural resources are the assets of every nation, enabling countries to be preferred when it comes to trade. Natural resource prevalence and environmental issues are not issues that are limited to geography; rather, it is a global challenge [1]. Using an economic lens, Reference [2] sees natural resources as the key to the progress of a country, describing natural resources as "factors of production provided by nature, which is, soils, forests, grassland, air, water, minerals, fuels, etc." In this vein, it is argued that the increasing depletion of natural resources is a severe threat to sustainable development.

Numerous methods are used to examine the impact of human and economic activities on environmental degradation. One such method is the ecological footprint, which first came to the surface in 1990, when it was described as "use of land and water for production of all resources consumed by humans and for eliminating the waste material generated by the population" [3]. In this process of production and consumption, the concept is generally used to examine the environmental situation, which is the outcome of these activities, and was earlier measured through CO_2 emissions [4]. Currently, the ecological footprint is generally used as an evaluation measure for environmental degradation [4]. A large part of the use of ecological footprints could be best described by the notion that an elaborative and comprehensive method is required to examine the effect of human activities on the environment. It is argued that natural resources positively affect ecological footprint and enable a country to manage its problem. For instance, in climate change or extreme heat, forestry significantly absorbs tons of carbon from spreading into the atmosphere. Further, trees reduce temperature and improve rainfall in the long run, which can help deal with water scarcity [3]. However, the excessive use of these natural resources, which is an "irreversible process", can pose grave challenges to human society and the prospect of environmental sustainability [5].

Apart from this, enhanced pressure on the ecological footprint results from greater demand for consumption and usage involved in attaining economic advancements, trade expansion, globalization, etc. With the increase in countries' desire to become highly globalized, the supply–demand tug of war has pushed countries to work together to minimize the supply–demand gap. Following this concept, many studies believe that globalization contributes to increasing pressure on the environment [6,7]. While explaining why globalization contributes to increasing pressure on the environment, they borrowed the idea of "race to bottom" earlier used by [8], which means that when host countries look for foreign direct investment, they relax their environmental regulations. Such relaxation generally allows countries to shift those businesses, which results in environmental challenges. Moreover, Reference [9] conducted a study on the relationship of globalization with the ecological footprint and found it to be a stressor, but the situation is different for the social aspect of globalization. They believed that the more the societies are interconnected and aware, the less the chances exist that stressors can play their role. On the other hand, the alternative view asserts that the emergence of globalization can have positive and negative effects on environmental changes [10]. In this regard, it is argued that developing countries often benefit from the learning curve of developed countries who have already honed their skills in confronting environmental challenges by developing green technology and processes.

Among the developing countries, the current use of natural resources in China is exceptional as the country alone has utilized 50% of global coal resources, and the impacts are quite visible in the forms of pollution and extreme weather situations [11]. Despite their importance, it has been observed that minimal work had been conducted in this specific area [3]. In this regard, Reference [12] shed light on natural resources in the Middle East and North Africa (MENA) region and drew attention to environmental and water shortage challenges. By describing the worsening situation of the region, Reference [12] used facts cited in work by the World Bank [13] and mentioned that the MENA region is one of the "poorest regions of the world in terms of renewable water and arable land", which is why the region is facing severe pollution challenges, environmental degradation and above all water shortage. Until recently, the situation has remained the same, and problems have further worsened despite the region's richness with other natural resources. Moreover, Reference [14] discussed this perspective differently; they argue that natural resource depletion can severally impact the environment. They further added that this can be managed if renewable energy resources are used instead of non-renewable resources. This can minimize ecological footprints and help maintain the natural resources that help deal with the climate change. In a somewhat similar manner, Reference [15] indicated the impact of natural resource (extraction) on environmental degradation. They suggested that in order to fulfill the increasing demand, the extraction of natural resources at higher rates decreases the bio capacity of the environment, which eventually results in ecological footprints.

Linking the use of resources with environmental burden, Reference [16] found the case of Thailand to be alarming. The ecological condition of Thailand was in surplus until the end of 1980s. Since 1990, the biocapacity surplus of Thailand has started to decline every year, indicating a continuous decline in environmental quality. In the year 2016, the ecological footprint of the country stood at 2.5 global hectares (gha). On the other hand, the biocapacity was 1.2. Hence, the biocapacity deficit was -1.3. In this regard, Reference [17] asserted that the higher pressure on natural resources has enhanced the ecological burden. Precisely, it is believed that the extension of oil palm and rubber plantations in the country has resulted in a rising ecological footprint in energy, forest and cropland. This increase in the area where the ecological footprint was previously greater than the biocapacity has amplified the stress on the utilization of resource use in the region. Given the increase in the globalization potential of Thailand, there is a need to identify the role of globalization in influencing environmental degradation. The association is crucial due to the rise in county's economic globalization and higher dependence on trade and foreign investments [18]. Additionally, recently, the study of [15] argued that economic growth in Thailand showed a decline in environmental degradation primarily but with subsequent deterioration due to the usage of outdated technologies and increased energy intensities.

In light of the above, it is wise to assert that economic growth, globalization and natural resources preserve the critical relationship with environmental degradation. Increasing human demands is creating stress, which is more than what the contemporary ecosystem can offer. Against this backdrop, the efforts to achieve equilibrium between demand and supply factors (economic growth, globalization, and natural resources) impact environment quality (ecological footprints). Though a plethora of studies available, it is clear that grounds are available to conduct research on the nexus of these three factors on environmental degradation. Critical review shows that limited research was conducted on natural research as a mechanism to minimize ecological footprints. Other grounds are also available to research ecological aspects of globalization. Keeping in view such possibilities, the present study examines the asymmetric impact of economic growth, globalization and natural resources on the ecological footprint. The present study utilizes a novel approach, nonlinear autoregressive distributed lag (NARDL) that was introduced by [19]. In particular, the present research investigates how positive and negative shocks of globalization and natural resources affect the ecological footprint in Thailand's economy. The outcomes of this study will provide a strong understanding of the relationship of globalization and natural resources with an ecological footprint in the context of Thailand.

Several investigations in the prevailing literature have emphasized the rising environmental challenges [20,21], mostly due to increasing globalization and industrial expansion [22–24]. The progress in country's development is desirable as it provides the government with the opportunity to serve the needs of the country better and sustain their future survival [25–27]. However, over time, with higher urbanization, trade and enhanced globalization, there is a continuous rise in environmental degradation resulting in increased consumption, water, air and land pollution [28,29], and enhanced exploitation of natural resources [30]. The traditional resource curse concept demonstrates that countries with abundant natural resources experience slower economic growth rates than countries that contain limited natural resources [31]. This is due to the negative effect that is transferred to other industries that subsequently pay for the prosperity of resource-associated industries [32]. Moreover, many natural resources, such as minerals and fuels as well as fishery and forestry, have also experienced the depletion that threatens the notion of sustainable development [3,33]. Additionally, there exist several adverse environmental impacts of natural resource extractions and consumption, especially fuels, such as coal, natural gas and petroleum, which have enhanced the levels of greenhouse gases into the atmosphere, thereby causes global warming [34].

Agreeing to the continuously growing notion of environmental deterioration, Reference [35] briefed on the major cause of this challenge, which in their opinion is the global

CO_2 emissions that are at times attributed to the economic growth. Contrary to this, there are voices that see the critical role of economic growth in a healthier environment. For instance, Reference [36] suggested that economic growth is the objective of any country, but it should be pursued by ensuring minimal environmental damage. Similarly, studies such as [4,12] also found a similar relationship between income and ecological footprint. This kind of relationship was also seen in the countries that are part of the Belt and Road initiative, where such a relationship is visible [37].

Contrary to this, there are studies that fail to find any such relationship. By way of illustration, Reference [38] work revealed that foreign direct investment (FDI), represented as a significant part of globalization, has no valid association with the ecological footprint. Similarly, the work of [39] is worthy of mention in this regard, which suggested that the positive or negative relationship between the income levels and ecological footprints is yet to reach a decisive point. The study also argued that such a relationship is absent or minimal in higher-income countries. On the basis of this, it is safe to mention that recently the biggest contributions to the environmental degradation are the developing countries. Furthermore, studies attempted to show the relationship between early phases of growth and environmental degradation. In this regard, one of the notable examples was provided by [40], who presented the case of Sub-Saharan countries and the consumption of energy. They revealed that countries like Botswana used less electricity before 2000 than today, so this consumption was raised by 3.8% at the Sub-Saharan African level. They believe the roots of this increasing consumption (mostly by non-renewable) could be traced from the improvement of gross domestic product (GDP) growth from 2.2% to 4.9%, gradually, from 2000 to 2017. In line with this, Reference [41] proposed that every country must take into consideration the need for balance between economic growth and environmental degradation. They believe that this can be achieved by the possibilities which push developed countries to control their revenue growth and also the developing countries to control their spread. The other possibility is through the domain of EKC link. It is believed that the increased levels of economic growth and globalization put pressure on manufacturing and consumption levels, leading to demand overshoot and an increase of the ecological footprint [42,43]. Similar concerns were traditionally raised in the environmental Kuznets curve (EKC) that suggests that rising income levels deteriorate initially but ultimately improved environmental quality [44–46]. However, there also exist concerns that the EKC only exists in the footprints of country's production and does not consider the globalization components, and is not reflected from import footprints [47]. This suggests that rich countries can easily improve their ecological footprint at the cost of poor countries' environment.

Hence, recognizing the potential threats of natural resource utilization, globalization and economic growth, many studies empirically analyzed the combined and specific impact of these variables on the environment. Among them, Reference [15] examined the impact of economic growth, financial advancements and energy utilization on environmental quality. To fulfill the objective, the authors utilized the measure of ecological footprints to identify environmental degradation in eleven newly industrialized nations from 1977 to 2013. The study stated that the role of economic progress is complementary to environmental degradation and sustainable development. Overall, the results suggested that energy consumption enhanced the ecological footprint in seven of the eleven economies. Likewise, financial development also decreases the ecological footprint of China and Malaysia but increases ecological overshoot in Singapore. As for economic growth, the outcomes found mixed results. For the economies of South Africa, Philippines, Mexico and Singapore, the results confirmed the existence of an inverted U-shaped EKC curve, suggesting that a rise in income initially amplified degradation but ultimately reduced it. On the other hand, the findings of Thailand, India, Turkey, China and South Korea suggested that a rise in economic growth declined ecological footprint primarily but subsequently deteriorated the environment due to the usage of outdated technologies and increased energy intensities.

Likewise, for a panel of MENA nations, Reference [12] examined the role of output growth in environmental degradation. Using the data of fifteen MENA economies from 1975 to 2007, the authors distributed the studied nations to oil-exporting and non-oil-exporting nations. The empirical investigation findings indicated an inverted U-shaped link between economic growth and ecological footprint in the MENA economies that export oil. For the case of non-oil-exporting nations, the study found the existence of a U-shaped EKC curve, suggesting that growth led to the reduction of ecological footprint, followed by a subsequent increase. Moreover, Reference [48] also analyzed the link between economic progress and environmental degradation in Qatar between 1980 and 2011. For this, the study adopted two major proxies of environmental degradation, i.e., carbon dioxide and ecological footprint. The outcomes found that the increase in economic growth decreased CO_2 and ultimately deteriorated environmental quality by increasing emission levels. On the other hand, the rise in economic growth degraded the environment by enhancing the ecological footprint but eventually improved environmental conditions with reduced pressure on the ecological footprint of Qatar.

In another study, Reference [49] analyzed European economies to study the link between economic growth and environmental degradation measured by ecological footprint. For this, the study used the data of fifteen European nations from 1980 to 2013. The study results found that an increase in economic growth reduced ecological footprint but ultimately amplified it. Moreover, in a mixed panel of 116 economies, Reference [47] also investigated the role of income levels in influencing the environment. Using the measure of ecological footprint to recognize environmental degradation, the study analyzed the validity of the EKC curve from 2004 to 2008. The study's findings reported that the existence of inverted U-curve association only existed in income and domestic production links. As for import footprint, the authors found that an increase in import led to enhanced ecological footprint monotonically. Similar to [49], Reference [50] also studied the role of economic growth and ecological footprint in Europe utilizing a panel of sixteen European economies between 1997 and 2014. The outcomes of the empirical results documented that a unit increase in economic growth is likely to raise the ecological footprint by 0.81%.

Likewise, Reference [51] also analyzed the growth-environment nexus in fourteen Asian economies using ecological footprint as an indicator of environmental degradation. The study results documented the validity of the EKC curve only in the economies of Nepal, Pakistan, India and Malaysia. However, for the rest of the Asian countries, including Thailand, the study found a significant positive relationship between growth and ecological footprint. Moreover, including natural resources in the environment-growth link, Reference [3] also examined the connection between output, natural resources and environmental degradation by adopting the proxy of ecological footprint to indicate climate downfall in Pakistan. Similar to [48], the study validated the presence of an inverted U-shaped EKC curve. As for natural resources, the results reported that natural resources increase the pressure on the environment by increasing the country's ecological footprint.

In another examination of natural resources and environment connection, Reference [52] analyzed the impact of natural resources in influencing the ecological footprint of China. Evaluating the data from 1980 to 2010, the results suggested the significant role of natural resources in enhancing ecological pressure in the Chinese economy. The outcomes reported that a rise in natural resource consumption carried a negative impact on ecological footprint leading to the enhanced ecological deficit by 66 times from 1983 to 2010. Likewise, in Thailand, Reference [17] investigated the impact of palm oil and rubber industries in affecting ecological footprint of the country. The results of the study found the significant role of the studied industries in affecting the ecological footprint of Thailand. The authors suggested that in order to attain the objectives of sustainability in rubber and palm oil industries, there remained the need to involve ecological footprint figures as the crucial indicator of sustainable growth. Focusing on oil resource, Reference [53] examined the oil and ecological footprint relationship in ten OPEC economies from 1977 to 2008. The results found the significant EKC link in the economies of Iraq, Nigeria, Kuwait, Algeria,

Venezuela and Qatar and reported the presence of an inverted U-Shaped association. Furthermore, the study found that an increase in oil consumption increased the ecological footprint in the considered economies.

Studying the link between globalization and environment, Reference [6] examined the connection between ecological footprint and king of fighters (KOF) index of globalization. In doing so, the authors evaluated the panel data of 171 economies for four diverse measures of ecological footprint, i.e., consumption, production, export and import footprints. The outcomes of the investigation reported the significant link of globalization on three measures of ecological footprint. Precisely, it is found that KOF index enhances environmental degradation by increasing import, export and consumption footprints in the studied economies. Moreover, Reference [54] also examined the impact of globalization on environmental degradation by utilizing the measure of ecological footprint. For this, the study gathered the data of 146 economies from 1981 to 2009. The findings supported the significant effect of overall globalization on the ecological footprint of export and import in the panel estimation. Additionally, the results indicated that the rise in social globalization decreases the footprints of production and consumption. On the other hand, an increase in social globalization is found to have a positive relationship with the footprints of export and import. As for economic globalization, the results found that economic globalization increases all types of ecological footprints. Lastly, the study failed to find the significant association between political globalization and the measures of ecological footprints.

Assessing the role of globalization in the context of EKC link, Reference [55] analyzed the association of economic growth and KOF index of globalization with an ecological footprint in South Asian economies from 1975 to 2017. The findings of the study validated the presence of EKC curve in the studied economies by reporting an inverted U-shaped association between economic growth and ecological footprint. Moreover, globalization tends to degrade environmental condition by enhancing the ecological footprint in South Asian countries. Moreover, distinguishing the environmental degradation into two measures of carbon and ecological footprint, Reference [10] investigated the influence of globalization on the Malaysian environment. In doing so, the authors used the data from 1971 to 2014 and reported the significant link of globalization in enhancing the carbon footprint of Malaysia. On the other hand, the study found that globalization persisted with an insignificant impact on the ecological footprint of Malaysia. In another recent study, Reference [42] also analyzed the impact of globalization on the ecological footprint of fifteen globalized economies from 1970 to 2017. They applied the innovative method of quantile-on-quantile regression, and the study reported the significant impact of globalization on the ecological footprint of the studied economies.

In addition, a study by Sharif et al. Reference [42] suggested that the ecological footprint has gained greater importance with time and needs to be investigated with some economic factors. Moreover, palm oil and rubber industries in Thailand have a greater effect on the ecological footprint and have a more significant impact on the country's economy. Thus, to fulfill these gaps and to consider the importance of this area, it motivates the researchers to examine the role of economic factors on ecological footprint. Thus, this research aims to examine the asymmetric effect of NAR and globalization (GLO) on the ecological footprint (EFP). Moreover, another aim is to test the environmental Kuznets curve in the Thailand economy.

2. Materials and Methods

This research examines the asymmetric effect of NAR and GLO on EFP. Moreover, it also tests the environmental Kuznets curve in the Thailand economy. Therefore, following this statement, the following equation is used for empirical estimation:

$$\ln \text{EFP}_t = f(\ln \text{GDP}_t, \ln \text{GDP}_t^2, \ln \text{NAR}_t, \ln \text{GLO}_t,) \qquad (1)$$

EFP = ecological footprint;
GDP = gross domestic product;

NAR = natural resources;
GLO = globalization.

The linear form of the above equation is as under:

$$\ln EFP_t = \beta + \beta_1 \ln GDP_t + \beta_2 \ln GDP_t^2 + \beta_3 \ln NAR_t + \beta_4 \ln GLO_t + \mu_t \quad (2)$$

lnEFP = logarithm of ecological footprint;
lnGDP = logarithm of gross domestic product;
lnNAR = logarithm of natural resources;
lnGLO = logarithm of globalization.

In econometric modeling, several approaches like ARDL, ECM or Granger causality are used to find the relationship between variables. The multiple regression analysis was applied since the independent variables are varied [56–61]. These approaches are generally used when the relationship between two or more variables needs to be checked, especially for the long run. One of the salient features of these approaches is their ability to take into account the asymmetric nature of the data. The data were collected from world development indicators and KOF index for globalization from 1970 to 2018. Contrary to this linear regression model is used for checking the linear relationship among variables, although they cannot check the variable nonlinear behavior. Taking the work on ARDL framework to a higher level, which is asymmetric ARDL co-integration approach: [19] based their efforts on the early contributions of [62,63], i.e., initial form ARDL framework. This newly developed approach captures short-term disturbances and also any asymmetries. This study is focused on exploring any such asymmetric effects of the independent variable on the dependent variable.

$$EFP_t = \alpha_0 + \alpha_1 GDP_t + \alpha_2 GDP_t^2 + \alpha_3 NAR_t^+ + \alpha_4 NAR_t^- + \alpha_5 GLO_t^+ + \alpha_6 GLO_t^- + \varepsilon_t \quad (3)$$

In this equation, the ecological footprint is denoted by EFP, whereas natural resources are represented with NAR, whilst globalization is denoted by GLO. Moreover, GDP and GDP² represent the gross domestic product and square of it, whereas the co-integrating vectors will be estimated by α (ranging from α1, α2, α3 to α6 in the equation). In addition, the partial positive and negative effect of focus variables (natural resources and globalization) on ecological footprint is also incorporated in Equation (3).

Considering Equation (2), as proposed by [19], the extended asymmetric ARDL model is shown as follows:

$$\Delta EFP_t = \beta_0 + \beta_1 EFP_{t-1} + \beta_2 GDP_{t-1} + \beta_3 GDP_{t-1}^2 + \beta_4 NAR_{t-1}^+ + \beta_5 NAR_{t-1}^- + \beta_6 GLO_{t-1}^+ + \beta_7 GLO_{t-1}^- + \sum_{i=1}^{m} \delta_{1i} \Delta EFP_{t-1} + \sum_{i=0}^{n} \delta_{2i} \Delta GDP_{t-i} + \sum_{i=0}^{n} \delta_{3i} \Delta GDP_{t-i}^2 + \sum_{i=0}^{p} \delta_{4i} \Delta NAR_{t-i}^- + \sum_{i=0}^{p} \delta_{5i} \Delta NAR_{t-i}^+ + \sum_{i=0}^{q} \delta_{6i} \Delta GLO_{t-i}^+ + \sum_{i=0}^{r} \delta_{7i} \Delta GLO_{t-i}^- + u_i \quad (4)$$

Equation (4) includes several lag orders, denoted by m, n, p, q and r. Similarly, NAR and GLO related effect of the disturbance, whether negative or positive on the EFP are reflected by β1, β2, β3, β4, and β5. Apart from them, Equation (4) also considers short term effects, which are represented by $\sum_{i=0}^{n} \delta_{2i}$, $\sum_{i=0}^{n} \delta_{3i}$, $\sum_{i=0}^{n} \delta_{4i}$, and $\sum_{i=0}^{n} \delta_{5i}$, respectively. Further, this is worth mentioning that a nonlinear long association among variables can also be examined using the NARDL approach.

The asymmetric ARDL model follows several steps; for instance, several tests like Augmented Dicky–Fuller and Phillips–Perron are performed as a first step. These tests will detail the stationarity of the variables, although it is not required when ARDL model is used. Researchers like [64–66] agree with the notion and argue that stationarity in a variable only acts as a hindrance if 1(2) series is present; otherwise, series such as 1(0), 1(1), or their mixture pose no threat to the application of ARDL model. In the backdrop of such a potential challenge, it is sane to examine these series for valid findings as the second step ordinary least square method is applied for the estimation of Equation (8). Aligned to this, the method of [67] was used for following SIC information criterion and general to specific approach in this regard. Lastly, co-integration was evaluated through the bound test, so the asymmetric ARDL model was used. This step made it a possibility to derive an asymmetric cumulative dynamic multiplier effect of percentage change in NAR_{t-1}^+, NAR_{t-1}^-, GLO_{t-1}^+, GLO_{t-1}^-, accordingly as shown as follows:

$$s_h^+(NAR) = \sum_{j=0}^h \frac{\partial EFP_{t+i}}{\partial NAR_{t-1}^+} \tag{5}$$

$$s_h^-(NAR) = \sum_{j=0}^h \frac{\partial EFP_{t+i}}{\partial NAR_{t-1}^-} \tag{6}$$

$$s_h^+(GLO) = \sum_{j=0}^h \frac{\partial EFP_{t+i}}{\partial GLO_{t-1}^+} \tag{7}$$

$$s_h^-(GLO) = \sum_{j=0}^h \frac{\partial EFP_{t+i}}{\partial GLO_{t-1}^-} \tag{8}$$

3. Results

In the initial step, the present research applied fundamental statistics, which is called descriptive statistics. The findings of descriptive statistics are reported in Table 1. It includes mean values along with minimum and maximum values for every variable opted in this research. Moreover, the table reported standard deviation, kurtosis, skewness and Jarque–Bera test to check the normality of the variables. The mean value shows the average value of the variable during the period, while standard deviation shows the deviation of the values from their mean. The findings confirm that the average value for all variables is positive. The skewness and kurtosis show the normality of the data. In addition, the present study utilized the Jarque–Bera test to assert the normality in the chose factors. The discoveries of the JB test assert the expulsion of the null hypothesis at a 1% level of criticalness, which suggests that each factor is non-linear. The results further assert that there implies nonlinearity in each selected factor [42,68,69].

Table 1. Descriptive statistics analysis.

Variable	EFP	GDP	GLO	NAR
Mean	1.714	3053.656	51.470	1.709
Minimum	0.955	929.091	32.444	0.562
Maximum	2.644	6128.658	69.129	3.785
Std. Dev.	0.587	1633.931	13.230	0.811
Skewness	0.084	0.274	−0.007	0.513
Kurtosis	1.448	1.768	1.391	2.373
Jarque-Bera	4.872	31.632	5.179	12.889
Probability	0.088	0.000	0.075	0.000

Source: authors' calculation.

There is an essential precondition of utilizing the ARDL bound testing methodology that the whole of the series of factors ought to be stationary at I(0) or I(1), nonetheless, not I(2). As appeared by Ouattara (2004), the disclosures of ARDL would be unacceptable if there is an I(2) factor presented in the studied model. Along these lines, it is vital to pick the stationarity of the dataset. Accordingly, the present investigation used two conventional unit root tests (for example, ADF and PP), and the results of the ADF and PP unit root are shown in Table 2. The outcomes showed that EFP, GDP, GLO and NAR demonstrate non-stationary conduct at a level and later changed into stationary at the first difference series. Moreover, the present investigation correspondingly utilized a basic structural break unit root test, for example [70], which imitates interruptions as clarified by [71]. Thinking about the issue of the break in the time plan, utilizing [70], the investigation additionally observed that all of the variables are stationary at I(1) as reported in Table 3. Along these lines, it is confirmed that the present examination is utilized the ARDL technique as all the selected variables are not I(2).

Table 2. Unit Root Test Analysis.

Variables	Unit Root Test (ADF)				Unit Root Test (PP)			
	I(0)		I(1)		I(0)		I(1)	
	C	C and T	C	C&T	C	C&T	C	C&T
EFP	0.410	0.367	−5.037 ***	−4.702 ***	0.366	0.375	−5.295 ***	−4.988 ***
GDP	−0.201	−0.183	−3.872 ***	−4.091 ***	−0.213	−0.237	−4.007 ***	−3.780 ***
GLO	−0.774	−0.734	−3.263 ***	−3.531 ***	−0.705	−0.744	−3.215 ***	−3.030 ***
NAR	0.546	0.410	−4.289 ***	−4.363 ***	0.261	0.283	−4.994 ***	−5.088 ***

Note: EFP represents the ecological footprint, GDP describes the per capita of gross domestic product, GLO explains the globalization index including social, political and economic globalization, and NAR represents the rents for natural resources. Moreover *** refer to the level of significance at 1%. Source: Authors' calculation.

Table 3. Unit root test on Zivot–Andrews trended structural break.

Variable	Level		1st Difference	
	T- Stat.	Time Break	T- Stat.	Time Break
EFP	−0.907 (1)	2007	−6.554 (1) ***	1997
GDP	−0.414 (1)	2015	−6.577 (1) ***	1984
GLO	0.845 (1)	2001	−6.238 (1) ***	1999
NAR	−1.506 (1)	2013	−9.063 (1) ***	2010

Note: parenthesis refer to lag order. *** refer to significance at 1% level. Source: authors' calculation.

Moreover, Reference [72] expressed that long-term affiliations concentrated on the best lag, and [73] likewise affirmed that utilizing an extra number of lags or taking a fewer lag could lose the most extreme imperious evidence of the model or might reason one-sided or biased estimations. Hence, sighted the status of perfect lags, the present investigation just 1 lag following the Schwarz info criteria (SIC). The discoveries of bound testing and nonlinear estimations are shown in Table 4. The outcome of F-statistics is greater than the tabulated values, which guarantees nonlinear long-term association among EFP, GDP, GLO and NAR in Thailand. Considering all the facts, the present examination pushes ahead to assess nonlinear ARDL coefficients.

Table 4. Bond test co-integration results.

Model	F-Stat.	Up. Bond	Low. Bond
ln EFP/(ln GDP, ln GDP^2, ln GLO_POS, ln GLO_NEG, ln NAR_POS, ln NAR_NEG)	64.583		
Critical Values			
0.10		4.50	1.70
0.05		5.40	2.20
0.01		6.90	2.80

Source: authors' calculation. Note: $p = o^+ = o^- = 0$, refer to combine null of no long-run relationship. The critical values are based on Narayan (2005).

After affirming the noteworthy nonlinear connection between EFP, GDP, GDP2, GLO and NAR in Thailand's economy, the present examination will continue towards long-run coefficients of our studied factors. The outcomes of long-run coefficients are reported in Table 5. The discoveries of NARDL affirmed that all factors altogether significant on the ecological footprint in Thailand. The outcomes further proposed that nonlinear and asymmetric association is found among EFP, GDP, GDP2, GLO and NAR in Thailand's economy. The outcomes also suggested that economic growth and squared of economic growth significantly impact the ecological footprint in Thailand. The results further suggested that due to the negative shocks of globalization, the ecological footprint is increased by 28.5%; however, due to the positive shocks of GLO, the EFP has also been increased by 10.3%. The trend of both the shocks is significant and positive; however, the magnitudes of both shocks are significantly different from each other, which suggested a nonlinear association between GLO and EFP in Thailand. On the other hand, the effect of NAR on the EFP is significant and positive. The negative shocks of NAR increase the EFP by 24.3%; however, the positive shocks of NAR increase the EFP by 40.3%. In this case, the signs of both shocks are positive, but again the sizes of coefficients are significantly different from each other, suggesting a presence of nonlinear connection between NAR and EFP in Thailand's economy.

Table 5. NARDL Approach for long-run asymmetric.

Variables	Coeff.	t-Stats	Prob.
ln GDP	0.375	4.092	0.000
ln GDP^2	−0.185	2.095	0.049
ln GLO_NEG	0.285	3.483	0.000
ln GLO_POS	0.103	4.094	0.000
ln NAR_NEG	0.243	3.968	0.000
ln NAR_POS	0.403	4.572	0.000

Dependent variable: ecological footprint. Source: authors' calculation.

This means that both globalization and natural resources are sources to increase the ecological footprint in Thailand. These findings are very rationale and justifiable as the consumption of natural resources and globalization, which mostly involved trade, increases the demand for natural resources, ultimately increasing the ecological footprint in a country. Moreover, the current study utilized nonlinear ARDL approach to test the environmental Kuznets curve in Thailand. The results suggested that economic growth is positive and significant; however, the square of economic growth is negative and significant, offering an inverted U-shape curve in Thailand. The results further confirm that, initially, the selected variables increase the ecological footprint, but after reaching a certain point, they started reducing the level of ecological footprint in Thailand.

Next, the discoveries of the diagnostic statistics of the NARDL technique are represented in Table 6. At this point, the criticalness estimation of LM and Breusch–Pagan–Godfrey are more noticeable than 0.100, which declares that the model is free from heteroscedasticity and serial correlation issues. Besides, the present examination has uncovered the p-estimation of the Ramsay RESET test, which is similarly more than 0.100, recommending that the current framework is sensibly specified. Finally, the present investigation points out the VIF estimation, which is 6.953, recommending no multicollinearity issue in the study's model.

Table 6. Diagnostic tests analysis.

Diagnostic Test	Problem	p-Value	Status
LM test	Serial Corr.	0.192	No Issue
BPagan-Godfrey	Hetero.	0.402	No Issue
Ramsey RESET test	Specification Err	0.731	No Issue
VIF	Multicoll.	6.953	No Issue

Source: authors' calculation.

In the final phase, the present research utilized the asymmetric Granger causality introduced by [74]. The present research has opted for asymmetric causality to investigate the causal connection between the positive and negative shocks globalization, natural resources and ecological footprint in Thailand's economy. The outcomes are shown in Table 7. The findings of asymmetric Granger causality confirm that negative shocks of GLO and EFP have a significant causal relationship with the negative shocks of GLO and EFP where the causality is running from the negative shocks of both variables to the negative shocks of other variables. In simple words, the findings confirm a significant bi-directional causal relationship between negative shocks of globalization and ecological footprint. On the other hand, the discoveries of asymmetric causality confirm that positive and negative shocks of NAR have a significant causal connection to the positive and negative shocks of EFP. However, the present study does not find any causal connection between positive and negative shocks of EFP to the positive and negative shocks of NAR in Thailand's economy.

Table 7. Asymmetric Granger causality analysis.

Null Hypothesis	Wald Test	Bstrap 1%	Bstrap 5%	Bstrap 10%
GLO^− does not Granger cause EFP^−	78.382 **	81.926	61.203	39.635
GLO^− does not Granger cause EFP^+	4.5832	45.160	35.205	25.948
GLO^+ does not Granger cause EFP^−	26.782	57.207	44.255	28.652
GLO^+ does not Granger cause EFP^+	59.391	130.870	103.732	89.953
EFP^− does not Granger cause GLO^−	38.582 **	40.948	32.148	24.452
EFP^− does not Granger cause GLO^+	41.582	77.102	63.682	50.538
EFP^+ does not Granger cause GLO^−	29.879	51.989	39.880	31.312
EFP^+ does not Granger cause GLO^−	18.116	50.685	41.546	32.404
NAR^− does not Granger cause EFP^−	127.520 ***	82.963	60.801	45.634
NAR^− does not Granger cause EFP^+	59.727 **	103.192	53.748	39.294
NAR^+ does not Granger cause EFP^−	177.547 ***	125.553	95.874	84.549
NAR^+ does not Granger cause EFP^+	295.091 ***	89.970	72.123	56.408
EFP^− does not Granger cause NAR^−	38.143	100.058	63.889	51.070
EFP^− does not Granger cause NAR^+	33.085	78.674	50.538	41.747
EFP^+ does not Granger cause NAR^−	1.447	39.837	24.397	7.900
EFP^+ does not Granger cause NAR^−	23.247	104.258	77.950	53.434

Note: ** and *** indicate statistical significance at 5% and 1% level, respectively. Critical values are obtained from 10,000 bootstrap replications. Source: authors' calculation.

4. Discussion and Conclusions

The present study investigated the asymmetric impact of natural resources and globalization on ecological footprint in the presence of EKC in Thailand. The study used annual time series data from 1970 to 2018. The findings confirm that the effect of globalization and natural resources are significant and nonlinear. These results align with Figge et al. [6], who also exposed that globalization and natural resources significantly affect the ecological footprint. However, the effect of negative shocks of globalization and natural resources is more dominant on the ecological footprint in Thailand than positive shocks of both variables. These results are also the same as Schandl et al. [28], who also examined that natural resources and globalization positively associate with the ecological footprint. Moreover, the present study has also tested the presence of EKC in Thailand, and the findings confirm the presence of an inverted U-shape curve in Thailand's economy. These results are also similar to Destek et al. [15], who also found the presence of an inverted U-shape curve in newly industrialized countries. On the other hand, the findings of asymmetric Granger causality confirm a bi-directional causal connection from negative shocks of globalization (ecological footprint) to the negative shocks of ecological footprint (globalization). This outcome is matched with the outcome of Charfeddine et al. [12] that also found bi-directional causality among globalization and ecological footprint. Moreover, the findings further suggested a unidirectional causal connection between natural resources and ecological footprint where causality runs from the positive and negative shocks of natural resources to the positive and negative shocks of ecological footprint. These outcomes are also in line with the output of Hassan et al. [3] that also found unidirectional causal relation among natural resources and ecological footprint.

Thus, the present study has concluded that the high level of natural resources and increasing level of globalization put a significant role on the ecological footprint in Thailand. In addition, as much as the natural resources hold by the country, the people of the country consume the resources faster and generate wastage. Moreover, globalization also forces people to use extra-ordinary resources to survive in the global market. Thus, this study suggested to the regulators that they should develop effective policies related to the effective usage of natural resources and positively respond to globalization affecting the environment.

Author Contributions: Conceptualization, H.I.H. and M.H.; methodology, H.I.H.; software, M.H.; validation, M.H., Z.D.-P. and K.S.-W.; formal analysis, F.K. and M.H.; investigation, Z.D.-P.; resources, K.S.-W.; data curation, K.S.-W.; writing—original draft preparation, H.I.H.; writing—review and editing, F.K.; visualization, Z.D.-P.; supervision, F.K.; project administration, H.I.H.; funding acquisition, F.K. All authors have read and agreed to the published version of the manuscript.

Funding: This research is funded by publication fund using PTJ code 12051, project code 9001103 under Universiti Putra Malaysia (UPM) and the project is also funded under the program of the Minister of Science and Higher Education titled "Regional Initiative of Excellence" in 2019–2022, project number 018/RID/2018/19, the amount of funding PLN 10 788 423,16.

Institutional Review Board Statement: Not applicable.

Informed Consent Statement: Not applicable.

Data Availability Statement: Data available in a publicly accessible repository that does not issue DOIs Publicly available datasets were analyzed in this study. This data can be found here: https://kof.ethz.ch/en/forecasts-and-indicators/indicators/kof-globalisation-index.html, assessed on 1 January 2021; https://databank.worldbank.org/source/world-development-indicators, assessed on 1 January 2021.

Acknowledgments: We would like to thank the editors and the anonymous referees of the journal for constructive comments and suggestions, which have significantly helped to improve the contents of the paper. The usual caveats apply.

Conflicts of Interest: The authors declare no conflict of interest.

References

1. Kapur, R. Natural Resources and Environmental Issues. *J. Ecosyst. Ecography* **2016**, *6*, 2–5. [CrossRef]
2. Johnson, B.; Villumsen, G. Environmental aspects of natural resource intensive development: The case of agriculture. *Innov. Dev.* **2017**, *8*, 167–188. [CrossRef]
3. Hassan, S.T.; Xia, E.; Khan, N.H.; Shah, S.M.A. Economic growth, natural resources, and ecological footprints: Evidence from Pakistan. *Environ. Sci. Pollut. Res.* **2019**, *26*, 2929–2938. [CrossRef]
4. Ulucak, R.; Bilgili, F. A reinvestigation of EKC model by ecological footprint measurement for high, middle and low income countries. *J. Clean. Prod.* **2018**, *188*, 144–157. [CrossRef]
5. Khan, I.; Hou, F.; Le, H.P. The impact of natural resources, energy consumption, and pollution growth on environmental quality: Fresh evidence from the United States of America. *Sci. Total Environ.* **2021**, *754*, 142222.
6. Figge, L.; Oebels, K.; Offermans, A. The effects of globalization on Ecological Footprints: An empirical analysis. *Environ. Dev. Sustain.* **2017**, *19*, 863–876. [CrossRef]
7. Twerefou, D.K.; Danso-Mensah, K.; Bokpin, G.A. The environmental effects of economic growth and globalization in Sub-Saharan Africa: A panel general method of moments approach. *Res. Int. Bus. Financ.* **2017**, *42*, 939–949. [CrossRef]
8. Zhang, J. Delivering environmentally sustainable economic growth: The case of China. *Asian Soc.* **2012**, 1–25. Available online: https://asiasociety.org/files/pdf/Delivering_Environmentally_Sustainable_Economic_Growth_Case_China.pdf (accessed on 1 January 2021).
9. Rudolph, A.; Figge, L. *How Does Globalization Affect Ecological Pressures? A Robust EMPIRICAL analysis Using the Ecological Footprint*; Publisher University of Heidelberg: Heidelberg, Germany, 2015; p. 599. [CrossRef]
10. Ahmed, Z.; Wang, Z.; Mahmood, F.; Hafeez, M.; Ali, N. Does globalization increase the ecological footprint? Empirical evidence from Malaysia. *Environ. Sci. Pollut. Res.* **2019**, *26*, 18565–18582. [CrossRef] [PubMed]
11. Zhang, B.; Meng, Z.; Zhang, L.; Sun, X.; Hayat, T.; Alsaedi, A.; Ahmad, B. Exergy-based systems account of national resource utilization: China 2012. *Resour. Conserv. Recycl.* **2018**, *132*, 324–338. [CrossRef]
12. Charfeddine, L.; Mrabet, Z. The impact of economic development and social-political factors on ecological footprint: A panel data analysis for 15 MENA countries. *Renew. Sustain. Energy Rev.* **2017**, *76*, 138–154. [CrossRef]
13. World Bank. 2002. The Annual Report. Washington, DC, USA. Available online: http://documents.worldbank.org/curated/en/379051468163155729/Main-report (accessed on 1 January 2021).

14. Sarkodie, S.A.; Strezov, V. A review on Environmental Kuznets Curve hypothesis using bibliometric and meta-analysis. *Sci. Total. Environ.* **2019**, *649*, 128–145. [CrossRef]
15. Destek, M.A.; Sarkodie, S.A. Investigation of environmental Kuznets curve for ecological footprint: The role of energy and financial development. *Sci. Total Environ.* **2019**, *650*, 2483–2489. [CrossRef]
16. Musikavong, C.; Gheewala, S.H. Assessing ecological footprints of products from the rubber industry and palm oil mills in Thailand. *J. Clean. Prod.* **2017**, *142*, 1148–1157. [CrossRef]
17. Musikavong, C.; Gheewala, S.H. Ecological footprint assessment towards eco-efficient oil palm and rubber plantations in Thailand. *J. Clean. Prod.* **2017**, *140*, 581–589. [CrossRef]
18. Chen, L.; Lombaerde, P.D. ASEAN between globalization and regionalization. *Asia Pac. Bus. Rev.* **2019**, *25*, 729–750. [CrossRef]
19. Shin, Y.; Yu, B.; Greenwood-Nimmo, M. Modelling asymmetric cointegration and dynamic multipliers in a nonlinear ARDL framework. In *Festschrift in Honor of Peter Schmidt*; Springer: New York, NY, USA, 2014; pp. 281–314.
20. Evans, J.P. 21st century climate change in the Middle East. *Clim. Chang.* **2009**, *92*, 417–432. [CrossRef]
21. Erlandson, J. Racing a Rising Tide: Global Warming, Rising Seas, and the Erosion of Human History. *J. Isl. Coast. Archaeol.* **2008**, *3*, 167–169. [CrossRef]
22. Weaver, R.H.; Jackson, A.; Lanigan, J.; Power, T.G.; Anderson, A.; Cox, A.E.; Eddy, L.; Parker, L.; Sano, Y.; Weybright, E. Health Behaviors at the Onset of the COVID-19 Pandemic. *Am. J. Heal. Behav.* **2021**, *45*, 44–61. [CrossRef]
23. Leman, M.A.; Claramita, M.; Rahayu, G.R. Predicting Factors on Modeling Health Behavior: A Systematic Review. *Am. J. Heal. Behav.* **2021**, *45*, 268–278. [CrossRef] [PubMed]
24. Habanabakize, T. The effect of economic growth and exchange rate on imports and exports: The south african post-2008 financial crisis case. *Int. J. Econ. Financ. Stud.* **2020**, *12*, 223–238. [CrossRef]
25. Haseeb, M.; Kot, S.; Hussain, H.I.; Kamarudin, F. The natural resources curse-economic growth hypotheses: Quantile–on–Quantile evidence from top Asian economies. *J. Clean. Prod.* **2021**, *279*, 123596. [CrossRef]
26. Hassan, A.; Meyer, D. Financial development–income inequality nexus in South Africa: A nonlinear analysis. *Int. J. Econ. Financ.* **2021**, *12*, 15–25.
27. Ślusarczyk, B.; Pypłacz, P. Industry 4.0 in Polish SMEs in the Aspect of Innovation Possibilities. *Int. J. Econ. Financ. Stud.* **2020**, *12*, 102–114.
28. Schandl, H.; Hatfield-Dodds, S.; Wiedmann, T.; Geschke, A.; Cai, Y.; West, J.; Newth, D.; Baynes, T.; Lenzen, M.; Owen, A. Decoupling global environmental pressure and economic growth: Scenarios for energy use, materials use and carbon emissions. *J. Clean. Prod.* **2016**, *132*, 45–56. [CrossRef]
29. Åhman, M.; Nilsson, L.J.; Johansson, B. Global climate policy and deep decarbonization of energy-intensive industries. *Clim. Policy* **2017**, *17*, 634–649. [CrossRef]
30. Rashid, A.; Irum, A.; Malik, I.A.; Ashraf, A.; Rongqiong, L.; Liu, G.; Ullah, H.; Ubaid Ali, M.; Yousaf, B. Ecological footprint of Rawalpindi; Pakistan's first footprint analysis from urbanization perspective. *J. Clean. Prod.* **2018**, *170*, 362–368. [CrossRef]
31. Ross, M.L. The political economy of the resource curse. *World Politics* **1999**, *51*, 297–322. [CrossRef]
32. Coxhead, I. A new resource curse? Impacts of China's boom on comparative advantage and resource dependence in Southeast Asia. *World Dev.* **2007**, *35*, 1099–1119. [CrossRef]
33. Aquilas, N.A.; Forgha, N.G.; Mobit, M.; Agbor, M.S. Natural Resources Depletion and Economic Growth: Implications and Prospect for Cameroon Economic Emergence by 2035. In *Natural Resource Endowment and the Fallacy of Development in Cameroon*; Publisher Langaa RPCIG: Bamenda, Cameroon, 2019; Volume 31.
34. Beer, T.; Grant, T.; Williams, D.; Watson, H. Fuel-cycle greenhouse gas emissions from alternative fuels in Australian heavy vehicles. *Atmos. Environ.* **2002**, *36*, 753–763. [CrossRef]
35. Balsalobre-lorente, D.; Shahbaz, M.; Roubaud, D.; Farhani, S. How economic growth, renewable electricity and natural resources contribute to CO2 emissions? *Energy Policy* **2018**, *113*, 356–367. [CrossRef]
36. Fakher, H. Investigating the determinant factors of environmental quality (based on ecological carbon footprint index). *Environ. Sci. Pollut. Res. Int.* **2019**, *26*, 10276–10291. [CrossRef] [PubMed]
37. Baloch, M.A.; Zhang, J.; Iqbal, K.; Iqbal, Z. The effect of financial development on ecological footprint in BRI countries: Evidence from panel data estimation. *Environ. Sci. Pollut. Res.* **2019**, *26*, 6199–6208. [CrossRef] [PubMed]
38. Solarin, S.A.; Al-mulali, U. Influence of foreign direct investment on indicators of environmental degradation. *Environ. Sci. Pollut. Res. Int.* **2018**, *25*, 24845–24859. [CrossRef]
39. Ozturk, I.; Al-Mulali, U.; Saboori, B. Investigating the environmental Kuznets curve hypothesis: The role of tourism and ecological footprint. *Environ. Sci. Pollut. Res.* **2016**, *23*, 1916–1928. [CrossRef] [PubMed]
40. Wang, J.; Dong, K. What drives environmental degradation? Evidence from 14 Sub-Saharan African countries. *Sci. Total Environ.* **2019**, *656*, 165–173. [CrossRef]
41. Mohiuddin, O.; Asumadu-Sarkodie, S.; Obaidullah, M. The relationship between carbon dioxide emissions, energy consumption, and GDP: A recent evidence from Pakistan. *Cogent Eng.* **2016**, *3*, 1–16. [CrossRef]
42. Sharif, A.; Afshan, S.; Qureshi, M.A. Idolization and ramification between globalization and ecological footprints: Evidence from quantile-on-quantile approach. *Environ. Sci. Pollut. Res. Int.* **2019**, *26*, 11191–11211. [CrossRef]
43. Hussain, H.I.; Abidin, I.S.Z.; Kamarulzaman, R.; Shawtari, F.A. The Impact of State Affiliated Directors on the Capital Structure Speed of Adjustment in an Emerging Market 2019. *Pol. J. Manag. Stud.* **2018**, *18*, 133–148.

44. Al-Mulali, U.; Weng-Wai, C.; Sheau-Ting, L.; Mohammed, A.H. Investigating the environmental Kuznets curve (EKC) hypothesis by utilizing the ecological footprint as an indicator of environmental degradation. *Ecol. Indic.* **2015**, *48*, 315–323. [CrossRef]
45. Dinda, S. Environmental Kuznets curve hypothesis: A survey. *Ecol. Econ.* **2004**, *49*, 431–455. [CrossRef]
46. Dinda, S. A theoretical basis for the environmental Kuznets curve. *Ecol. Econ.* **2005**, *53*, 403–413. [CrossRef]
47. Aşıcı, A.A.; Acar, S. Does income growth relocate ecological footprint? *Ecol. Indic.* **2016**, *61*, 707–714. [CrossRef]
48. Mrabet, Z.; Alsamara, M. Testing the Kuznets Curve hypothesis for Qatar: A comparison between carbon dioxide and ecological footprint. *Renew. Sustain. Energy Rev.* **2017**, *70*, 1366–1375. [CrossRef]
49. Destek, M.A.; Ulucak, R.; Dogan, E. Analyzing the environmental Kuznets curve for the EU countries: The role of ecological footprint. *Environ. Sci. Pollut. Res.* **2018**, *25*, 29387–29396. [CrossRef] [PubMed]
50. Alola, A.A.; Bekun, F.V.; Sarkodie, S.A. Dynamic impact of trade policy, economic growth, fertility rate, renewable and non-renewable energy consumption on ecological footprint in Europe. *Sci. Total Environ.* **2019**, *685*, 702–709. [CrossRef]
51. Uddin, G.A.; Alam, K.; Gow, J. Ecological and economic growth interdependency in the Asian economies: An empirical analysis. *Environ. Sci. Pollut. Res.* **2019**, *26*, 13159–13172. [CrossRef]
52. Gao, J.; Tian, M. Analysis of over-consumption of natural resources and the ecological trade deficit in China based on ecological footprints. *Ecol. Indic.* **2016**, *61*, 899–904. [CrossRef]
53. Saboori, B.; Al-Mulali, U.; Bin Baba, M.; Mohammed, A.H. Oil-induced environmental Kuznets curve in organization of petroleum exporting countries (OPEC). *Int. J. Green Energy* **2016**, *13*, 408–416. [CrossRef]
54. Rudolph, A.; Figge, L. Determinants of Ecological Footprints: What is the role of globalization? *Ecol. Indic.* **2017**, *81*, 348–361. [CrossRef]
55. Sabir, S.; Gorus, M.S. The impact of globalization on ecological footprint: Empirical evidence from the South Asian countries. *Environ. Sci. Pollut. Res.* **2019**, *26*, 33387–33398. [CrossRef]
56. Hussain, H.I.; Abidin, I.S.Z.; Ali, A.; Kamarudin, F. Debt Maturity and Family Related Directors: Evidence from a Developing Market. *Pol. J. Manag. Stud.* **2018**, *18*, 118–134. [CrossRef]
57. Shafai, N.A.; Nassir, A.M.; Kamarudin, F.; Rahim, N.A.; Ahmad, N.H. Dynamic Panel Model of Dividend Policies: Malaysian Perspective. *Contemp. Econ.* **2019**, *13*, 239–252.
58. Martín-Moya, R.; Ruiz-Montero, P.J.; García, E.R.; Leeson, G. Psychological and environmental factors for older adults to exercise: A systematic review. *Rev. Psicol. Deporte* **2020**, *29*, 93–104.
59. Zhang, Y.; Zhang, L.; Zhang, M. The Influence of Organizational Support and Citizen Behavior on Job Performance of New Energy Enterprises: Mediating Effects of Psychological Capital. *Revista de Psicología del Deporte. J. Sport Psychol.* **2020**, *29*, 210–220.
60. Yang, Z. An Empirical Analysis of the Relationship between Self-efficacy, Motivation, and Team Performance of High-level Basketball Players. *Revista de Psicología del Deporte. J. Sport Psychol.* **2020**, *29*, 221–231.
61. Butler, A.E.; Battista, K.; Leatherdale, S.T.; Meyer, S.B.; Elliott, S.J.; Majowicz, S.E. Environmental Factors of Youth Milk and Milk Alternative Consumption. *Am. J. Health Behav.* **2020**, *44*, 666–680. [CrossRef]
62. Szczepańska-Woszczyna, K.; Kurowska-Pysz, J. Sustainable business development through leadership in SMEs. *Eng. Manag. Prod. Serv.* **2016**, *8*, 57–69. [CrossRef]
63. Székely, S.; Csata, Z.; Cioca, L.-L.; Benedek, A. Industrial marketing 4.0-upgrading the industrial costumers' path to the digital economy. *Pol. J. Manag. Stud.* **2020**, *22*, 535–548.
64. Kurowska-Pysz, J.; Wróblewski, Ł.; Szczepańska-Woszczyna, K. Identification and assessment of barriers to the development of cross-border cooperation. In *Innovation Management and Education Excellence through Vision 2020, Proceedings of the 31st International Business Information Management Association Conference, Milan, Italy, 25–26April 2018*; Publisher International Business Information Management Association: King of Prussia, PA, USA, 2018; pp. 3317–3327.
65. Hussain, H.I.; Hadi ARAMohamed-Isa, A.; Salem, M.A.; Kamarudin, F.; Jabarullah, N.H. Adjustment to Target Debt Maturity and Equity Mispricing: Evidence from Asia Pacific. *Polish J. Manag. Stud.* **2018**, *17*, 87–100. [CrossRef]
66. Hussain, H.I.; Ali, A.; Thaker, H.M.T.; Ali, M. Firm Performance and Family Related Directors: Empirical Evidence from an Emerging Market. *Contemp. Econ.* **2019**, *13*, 187–204.
67. Katrakilidis, C.; Trachanas, E. What drives housing price dynamics in Greece: New evidence from asymmetric ARDL cointegration. *Econ. Model.* **2012**, *29*, 1064–1069. [CrossRef]
68. Mishra, S.; Sharif, A.; Khuntia, S.; Meo, S.A.; Khan, S.A.R. Does oil prices impede Islamic stock indices? Fresh insights from wavelet-based quantile-on-quantile approach. *Resour. Policy* **2019**, *62*, 292–304. [CrossRef]
69. Raza, S.A.; Shah, N.; Sharif, A. Time frequency relationship between energy consumption, economic growth and environmental degradation in the United States: Evidence from transportation sector. *Energy* **2019**, *173*, 706–720. [CrossRef]
70. Zivot, E.; Andrews, D.W.K. Further evidence of the great crush 2019, the oil price shock and the unit root hypothesis. *J. Bus. Econ. Stat.* **1992**, *10*, 251–270.
71. Perron, P. The great crash, the oil price shock, and the unit root hypothesis. *Economet. J. Econ. Soc.* **1989**, *57*, 1361–1401. [CrossRef]
72. Sharif, A.; Afshan, S.; Khan, B.S. Does democracy embolden economic growth in Pakistan? Evidence from ARDL bound testing and rolling window. *Int. J. Econ. Bus. Res.* **2018**, *15*, 180. [CrossRef]
73. Raza, S.A.; Jawaid, S.T.; Afshan, S.; Karim, M.Z.A. Is stock market sensitive to foreign capital inflows and economic growth? Evidence from Pakistan. *J. Chin. Econ. Foreign Trade Stud.* **2015**, *8*, 142–164. [CrossRef]
74. Hatemi-j, A. Asymmetric causality tests with an application. *Empir. Econ.* **2015**, *43*, 447–456. [CrossRef]

Article

A Modification of Palm Waste Lignocellulosic Materials into Biographite Using Iron and Nickel Catalyst

Noor Hafidzah Jabarullah [1,*], Afiqah Samsul Kamal [2] and Rapidah Othman [2]

1. Malaysian Institute of Aviation Technology, University Kuala Lumpur, Sepang 43900, Malaysia
2. Chemical Engineering Section, Malaysian Institute of Chemical and Bioengineering, University Kuala Lumpur, Melaka 78000, Malaysia; afiqah.samsul30@s.unikl.edu.my (A.S.K.); rapidah@unikl.edu.my (R.O.)
* Correspondence: nhafidzah@unikl.edu.my

Abstract: This paper presents an alternative way to maximize the utilization of palm waste by implementing a green approach to modify lignocellulosic materials into a highly crystalline biographite. A bio-graphite structure was successfully synthesized by converting lignocellulosic materials via a simple method using palm kernel shell (PKS) as a carbon precursor. This involved the direct impregnation of a catalyst into raw material followed by a thermal treatment. The structural transformation of the carbon was observed to be significantly altered by employing different types of catalysts and varying thermal treatment temperatures. Both XRD and Raman spectroscopy confirmed that the microstructural alteration occurred in the carbon structure of the sample prepared at 800 and 1000 °C using iron, nickel or the hybrid of iron-nickel catalysts. The XRD pattern revealed a high degree of graphitization for the sample prepared at 1000 °C, and it was evident that iron was the most active graphitization catalyst. The presence of an intensified peak was observed at 2θ = 26.5°, reflecting the formation of a highly ordered graphitic structure as a result of the interaction between the iron catalyst and the thermal treatment process at 1000 °C. The XRD observation was further supported by the Raman spectrum in which PKS-Fe1000 showed a lower defect structure associated with the presence of a significant amount of graphitic structure, as a low value of (I_d/I_g) ratio was reported. An HRTEM image showed a well-defined lattice fringe seen on the structure for PKS-Fe1000; meanwhile, a disordered microstructure was observed for the control sample, indicating that successful structural modification was achieved with the aid of the catalyst. Further analysis from BET found that the PKS-Fe1000 developed a surface area of 202.932 m2/g with a pore volume of 0.208 cm3/g. An overall successful modification from palm waste into graphitic material was achieved. Thus, this study will help those involved in waste management to evaluate the possibility of a sustainable process for the generation of graphite material from palm waste. It can be concluded that palm waste is a potential source of production for graphite material through the adoption of the proposed waste management process.

Keywords: palm kernel shell; graphite; catalytic graphitization; biomass waste

Citation: Jabarullah, N.H.; Kamal, A.S.; Othman, R. A Modification of Palm Waste Lignocellulosic Materials into Biographite Using Iron and Nickel Catalyst. *Processes* **2021**, *9*, 1079. https://doi.org/10.3390/pr9061079

Academic Editor: Lucian-Ionel Cioca

Received: 19 May 2021
Accepted: 16 June 2021
Published: 21 June 2021

Publisher's Note: MDPI stays neutral with regard to jurisdictional claims in published maps and institutional affiliations.

Copyright: © 2021 by the authors. Licensee MDPI, Basel, Switzerland. This article is an open access article distributed under the terms and conditions of the Creative Commons Attribution (CC BY) license (https://creativecommons.org/licenses/by/4.0/).

1. Introduction

Malaysia and Indonesia are the largest producers of palm oil, contributing more than 80% of the global annual yield [1]. Oil palm is among the best known and most extensively cultivated plant families in Malaysia, as compared to another commodities [2]. Various products are derived from oil palm, such as in the food, cosmetic, animal feed and pharmaceutical industries [3]. However, increases in the demand for oil palm products leads to an increase in the waste produced annually [4]. A huge amount of biomass wastes produced from the palm oil industry comprises palm kernel shell (PKS), empty fruit bunch (EFB) and palm mesocarp fibre (PMF) [1,5].

Current European Union waste management directives promote the prevention of waste and the application of waste management hierarchy [6]. Sustainable waste management is an essential step to overcoming the waste generation issue [7–9]. The hierarchy of waste management includes prevention, reuse, recycling, recovery and disposal [10]. The utilization of waste plays an important role in the transition to a circular economy as waste is reused to produce other valuable biomaterials that indirectly enhance economic growth [2,11,12]. This is related to the waste to wealth concept.

Lignocellulosic biomass waste is an ideal source of cellulose-based natural fibre and a good source of biomaterial [2]. It has received a large amount of attention due to its potential to be utilized in a wide range of applications, such as the development of biocomposites, power generation, paper production, biofertilizers, super-capacitors and water purification [2,5,13,14]. Lignocellulosic biomass is a natural, non-toxic, abundant, sustainable and renewable material, making it a good candidate for a wide range of applications [15–17].

Lignocellulosic biomass is naturally porous with a hierarchical structure and large surface area [18]. The porous structure is beneficial as it provides an ion transport pathway; it can also be easily combined with other materials to form a hybrid material that exhibits enhanced material electrochemical properties [19]. Lignocellulosic is also a carbon-rich material with heteroatom-doping characteristics, making it attractive for use as a substitute of conventional carbon material [18]. Biomass-derived carbon materials have given rise to extensive interest due to their biodiversity, unique microstructure and good conductivity [20–22].

Today, research has shifted toward abundantly available and low-cost materials, such as lignocellulosic waste [23]. Initiatives of producing carbon material from biomass waste will subsequently help in reducing environmental issues and maximizing utilization [24]. However, challenges occur as raw biomass is made up of many different macromolecules, small molecules and inorganic component, such as cellulose, hemicellulose, lignin and tannins [25]. Different components exhibit different thermal decomposition and chemical properties, which make the synthesis process difficult as compared to purified derivatives, such as cellulose [25].

Palm waste, such as palm kernel shell (PKS) waste, is currently being utilized as biofuel for steam generation as it has a higher heating rate than other palm wastes. Additionally, its small size, ease of handling and limited biological activity due to its lower moisture content make PKS attractive for use in the production of activated carbon, polymer biocomposite and as concrete pavement [26]. One strategy to fully utilize this biomass waste is using it as a main feedstock for the production of carbon material. Previously, many researchers' works have investigated different raw materials, such as sawdust, saccharides, corn stalk, green tea waste, coconut coir and coconut stalk waste as raw materials for carbon material production.

Carbon materials have raised great interest in a wide range of applications. This interest has resulted from its diverse beneficial properties. Properties such as high chemical stability, good electrical conductivity and high thermal stability have increased interest in many areas. Carbon materials have been widely used in filtration, separation technology, electrodes and energy devices. There are several types of carbon material that are prominent among researchers, including carbon nanotubes, graphene and graphene oxide.

Among the carbon materials that are widely investigated is graphite. Graphite is one form of carbon allotrope that has attracted tremendous interest due to its mechanical, electrical and thermal properties [27,28]. The metallic (high electrical and thermal conductivities) and non-metallic properties (high thermal resistance, inertness and lubricity) of graphite make it an essential material for refractories, automobiles, lithium-ion batteries, fuel cells, solar cells and graphene production [29–31].

The common method in graphite manufacturing is known as the Ancheson process [30]. The process involved the thermal heating of amorphous carbon up to 3000 °C. The energy and time consumption is tremendously high, as it involves about a 2 week processing period, leading to high production costs [30]. In a later development, graphite was also prepared using other methods, such as arc discharge, laser ablation, chemical vapour deposition (CVD) and solvothermal synthesis [32,33]. However, these methods require extreme condition, such as a high temperature, high-energy consumption and special and expensive equipment [33,34]. In some cases, it produces graphite with a low specific surface area, which is not favourable [34].

Catalytic graphitization, on the other hand, is a heat treatment process involving a metal catalyst to enhance the process [33]. Unlike the other methods mentioned above, catalytic graphitization involves moderate process conditions with a relatively low temperature and energy consumption. It utilizes solid feedstock, which is an attractive option due to its easy handling and transport. Therefore, a feasible method, such as catalytic graphitization using lignocellulosic materials as a carbon precursor, have become an attractive option [32,35,36].

Catalytic graphitization is the process of transforming amorphous carbon into a well-ordered graphitic structure by a heat treatment in the presence of metals or minerals. Preovious researchers have reported on the utilization of transition metals such as Fe, Co, Ni, Mg or metalloid elements, during graphitization processes using a variety of carbon precursors, ranging from coal wastes to biomass materials [25,37–41]. Among them, Fe, Ni and Co have been found to be effective catalysts in graphite production [41]. Among the most widely investigated catalyst is iron, as it was reported as the most effective catalyst in a number of research works [25,32,42–44]. Iron is an attractive metal due to its toxicity, magnetization and cost effectiveness [44]. Previous researchers have studied the effect of different nitrate salts of copper, nickel, cobalt and iron on graphitization and concluded that the iron nanoparticle was the most active catalyst [43]. Qiangu et al. (2018) also reported a comparable finding that iron was the most active catalyst [45].

Based on the decision to take advantage of lignocellulosic waste and focusing on transforming toward sustainable economy, it is vital to study the optimization of production conditions [2,5,6]. Although catalytic graphitization using palm kernel shell has been reported previously [46,47], these studies incorporated PKS with supporting materials, such as paraffin and petroleum coke, to form an activated carbon. It is very important to note that the structures of activated carbon and graphite are significantly different. The latter has a higher electrical and thermal conductivity with a lower surface area than the former. In our study, we successfully transformed the amorphous carbon structure of PKS into highly graphitic carbon similar to natural graphite using a heat treatment at 800 and 1000 °C with the presence of Fe, Ni and the hybrid Fe-Ni catalyst. We designate the highly graphitic structure of PKS as biographite.

2. Materials and Methods

2.1. Sample Preparation

Palm kernel shell (PKS) was collected from Bell KSL Kilang Sawit Linggi, Negeri Sembilan. The raw materials were firstly air-dried for 24 h to remove excess moisture before being cut and chopped into smaller sizes. Approximately 150 g of the material was washed with hot deionized water at 80 °C to remove impurities. Later, the material was oven dried at 70 °C for 24 h. Then, the washed material was ground and sieved to achieve a uniform size of 206 µm. The raw material was divided into three parts of 50 g each.

2.2. Catalyst Preparation

An amount of 1 M aqueous solution of iron (III) nitrate nonahydrate was prepared by mixing 40.399 g (10 wt.%) iron (III) nitrate nonahydrate in 100 mL deionized water at an ambient temperature of 25 °C. Nickel (II) nitrate hexahydrate solution was prepared by adding 29.079 g (10 wt.%) to 100 mL deionized water at an ambient temperature of 25 °C. Meanwhile, the hybrid ferum and nickel catalyst comprised a mixture of 1:1 part 40.399 g (10 wt.%) iron (III) nitrate nonahydrate and nickel (II) nitrate hexahydrate 29.079 g (10 wt.%) with 100 mL deionized water at an ambient temperature of 25 °C.

2.3. Bio-Graphite Preparation

An amount of 50 g of PKS sample was immersed in 100 mL of 1.0 M aqueous solutions of iron (II) nitrate hexahydrate (29.079 g) and stirred for 48 h. The samples were then filtered and dried at 80 °C. The sample was divided into two parts for graphitization at 1000 and 800 °C, respectively. The sample was placed in a quartz holder and inserted into a Carbolite 1800C Tube Furnace/Model CTF 18/300. The samples underwent heat treatment with a heating rate of 5 °C/min in a nitrogen atmosphere. After the graphitization, the residual metal catalyst was removed by stirring in 1.0 M HCl for 24 h followed by washing it using deionized water. Then, it was dried in an oven at a temperature of 80 °C overnight. The samples were denoted as PKS Fe-1000 and PKS Fe-800. The steps were repeated using an aqueous solution of nickel (II) nitrate hexahydrate and a hybrid of iron (III) nitrate nonahydrate and nickel (II) nitrate hexahydrate. Samples were denoted as PKS Ni-800, PKS Ni-1000, PKS Fe Ni-800 and PKS Fe Ni-1000. A reference control sample was also prepared by directly inserted PKS raw material into a tube furnace at 1000 °C without the addition of any catalyst, and the sample was denoted as PKS control-1000.

2.4. Characterization

X-ray diffraction (XRD) patterns were obtained from $2\theta = 2$ to $90°$ in a PAN analytic X'Pert Pro diffractometer using Cu Kα radiation ($\lambda = 1.5406$ Å, 45 kV, 40 mA), used for the identification of minerals. Raman spectra were recorded using the Renishaw Micro-Raman system 2000 with He-Ne laser excitation $\lambda = 632$. An FEI Tecnai G2 20 S-twin transmission electron microscope ((HR)TEM, FEI), operated at 200 kV and equipped with a filed-emission gun, supplied information about the structure of carbon materials. The surface area and pore size measurements were carried out by N_2 adsorption/desorption isotherms using Micromeritics2360, ASAP2020.

3. Results and Discussion

3.1. Crystallographic Properties

The XRD patterns in the wide angle region, 2–90°, permitted an evaluation of the graphitic nature of the synthesized biographite. The analysis was carried out to characterize the crystallinity of the composites [48]. PKS samples were introduced into Fe, Ni and FeNi-catalysts, respectively, and underwent heat treatment at a temperature of 800 and 1000 °C. The XRD patterns for all samples prepared under different conditions are shown in Figure 1. For the optimal synthesis condition of lignocellulosic waste modification, two main factors, including heat treatment temperature and type of catalyst, were among the important aspects. The effect of the temperature and catalyst were determined by XRD patterns.

Figure 1 shows the XRD patterns for each sample at 800 and 1000 °C with 10 wt.% catalyst loading, and the control sample. Most of the samples exhibited the presence of a sharp peak located at $2\theta \sim 26°$, corresponding to the (002) plane of graphite [40,49–51]. This confirms the transformation of PKS waste to a bio-graphite structure. The samples prepared using iron, nickel and a hybrid iron-nickel catalyst also exhibited diffraction peaks at $2\theta \sim 44°$, 50° and 59.98°, corresponding to (101), (102) and (103) diffractions of graphite frameworks, respectively.

It can be clearly observed that the sample prepared at a higher temperature showed strong reflection at $2\theta \sim 26°$, which is associated with a strong degree of graphitization, regardless of the type of catalyst used. The peak was significantly intensified with an increase in the graphitization temperature. These results are similar to those reported by previous research works [37,40,52–54]. The observations were also comparable to the work carried out by Yang et al. (2019), which utilized sucrose as a carbon precursor and successfully produced a well-ordered graphite structure at higher temperatures [55].

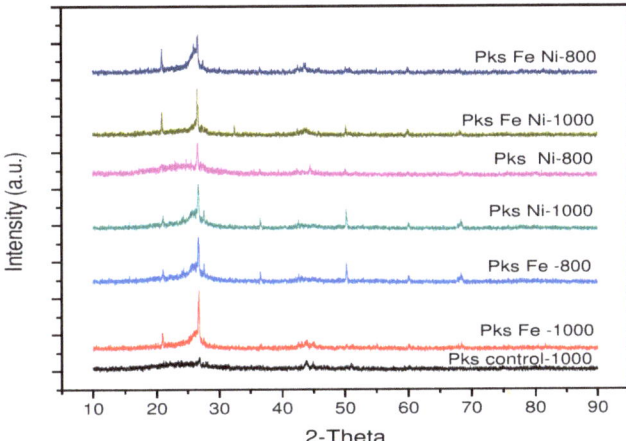

Figure 1. XRD pattern for PKS sample prepared with different temperatures and catalysts.

To further discuss the effect of catalyst selection, XRD profiles for PKS samples prepared with different catalysts at 1000 °C were compared. A significant sharp and narrow diffraction peak obtained for the sample prepared using an iron catalyst at $2\theta \sim 26°$ indicates that a high degree of crystallinity of graphitized carbon can be obtained by using iron as compared to nickel and a hybrid iron-nickel catalyst. The PKS control-1000 sample, on the other hand, showed almost no visible graphite peak. The PKS control sample was prepared by undergoing heat treatment at 1000 °C without any catalyst. This proved that PKS waste with the absence of a catalyst is not graphitizable at 1000 °C.

From the XRD pattern, it can be deduced that the degree of graphitization increased with the graphitization temperature and the application of an iron catalyst. Therefore, the value of d_{002} spacing was calculated according to Bragg's equation, as shown in Table 1 [56]. The value of d_{002} spacing was observed in the range between 0.3325 and 0.3351 nm, close to the value of graphite (0.3354 nm) and less than 0.344 nm for the disordered carbon material. Of all the catalysts, PKS Fe-1000 (0.3351 nm) was the nearest to 0.3354 nm, corresponding to the ideal graphite and suggesting that an ordered carbon framework was achieved [57]. This suggests that the structure of PKS Fe-1000 was altered and transformed into graphite. Numerous previous studies have modified biomass into graphite, and the d-spacing data were mostly close to graphite. Xiangyang et al. (2016) used wheat stalk as a carbon precursor and reported d-spacing of 0.3362 nm, and Jujiao et al. (2016) reported 0.34 nm d-spacing using chitosan [58]. Other findings show d-spacing between 0.337 and 0.346 using various raw materials [24,28,44,55,59].

Table 1. d-spacing data of all graphitic carbon samples.

Sample	d_{002} nm
PKS Fe-800	0.3348
PKS Fe-1000	0.3351
PKS Ni-800	0.3325
PKS Ni-1000	0.3327
PKS FeNi-800	0.3343
PKS FeNi-1000	0.3344
PKS control-1000	0.3391
Graphite commercial	0.3354

The structural parameters of this sample were further deduced from the XRD profile via peak fitting. Crystallite sizes along the c-axis, Lc and a-axis La were deduced by means of Scherrer's equation applied to the (002) and (101) diffraction peaks [60]. The d_{002} values for the control sample were larger, 0.3391 nm, suggesting that it still has a turbostatic carbon structure.

The crystallographic structure modification of the PKS Fe-1000 sample was further detected by Raman spectra. Raman spectra can identify the presence of graphite or a disordered amorphous structure in the sample [45]. For the graphitic carbon material, two strong resonance peaks at 1580 and 1350 cm^{-1} were observed. A peak at 1580 cm^{-1} represents the vibration in an ideal graphite lattice (G) band [30,52,60]. Meanwhile, the D (1350 cm^{-1}) band appeared with the increase in structural defects due to imperfections or loss of hexagonal symmetry [45,61]. The peak intensity ratios of the G and D1 bands are indicators of the defect density of carbon material [52]. By means of Raman spectroscopy, it was possible to analyse the degree of structural organization of the graphitic carbon sample. The relative intensity ratio between the D and G bands (I_d/I_g) and width at half maximum of the G bands (Δv_G) reflected the degree of graphitization. Low values of I_d/I_g and Δv_G parameters clearly indicate a high degree of graphitization [50]. Raman spectra forPKS Fe-1000 sample was compared with commercial graphite and the PKS control sample in Figure 2.

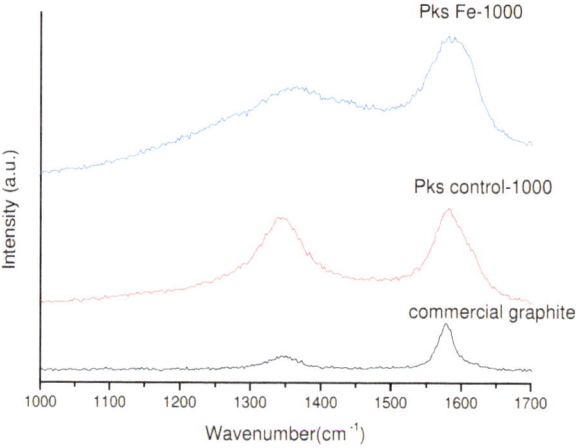

Figure 2. Raman spectra for Pks Fe-1000, PKS control and commercial graphite.

A higher D/G peak intensity (I_d/I_g) ratio reveals more structural defects in carbon material. As reported in Table 2, the I_d/I_g values for PKS Fe-1000, commercial graphite and PKS control-1000 were 0.985, 0.576 and 1.632, respectively. The I_d/I_g values were similar, as reported by a previous researcher, ranging between 0.84 and 0.98 I_d/I_g, with the utilization of lignin as a carbon source [20]. This further proved that thermal treatment at 1000 °C and an iron catalyst successfully induced a high degree of graphitization into the palm kernel shell (PKS) waste. The Δv_G parameter and I_d/I_g value showed a descending order value for the PKS control, PKS Fe-1000 and commercial graphite. Lower Δv_G and I_d/I_g values for PKS Fe-1000 showed that the successful alteration of the amorphous material was achieved. The Raman spectra data are comparable with the XRD findings. In addition, the control sample showed the highest Id/Ig value of 1.63 compared to PKS Fe-1000 and commercial graphite, suggesting the presence of a turbostratic structure. Both analyses from the XRD and Raman suggested that the graphitic structure of the PKS waste was remarkably produced with the aid of an iron catalyst and thermal treatment at 1000 °C.

Table 2. Physical properties of synthesized graphitic carbon, commercial graphite and control sample.

Sample	Structural Characteristic			Raman Parameter		Textural Properties		
	d_{002} (nm)	Lc (nm)	La (nm)	I_D/I_G	Δv_G (cm^{-1})	S_{BET} (m$^2 \cdot$g^{-1})	Pore Volume (cm$^3 \cdot$g^{-1})	Pore Size (Å)
Pks control	0.3391	6.78	60.14	1.631	92.94	17.014	0.017	37.251
Commercial graphite	0.3354	43.47	37.25	0.576	33.66	0.3748	0.014	1592.1
Pks Fe-1000	0.3351	62.00	11.12	0.985	83.40	202.93	0.208	41.067

3.2. Morphology and Pore Structure

To probe porosity, a nitrogen sorption–desorption isotherm was collected for PKS Fe-1000, PKS control and commercial graphite, as depicted in Figure 3. The Brunauer–Emmett–Teller surface area (SABET) of the sample is 202.932 m^2/g. The value was found to be close to a study conducted by Thompson (220 m^2/g), as biomass waste was also utilized in his research [25]. The pore size of the sample is 4.107 nm, which satisfied the range of pore size for mesoporous material (2–50 nm) and pore volume of 0.208 cm^3/g.

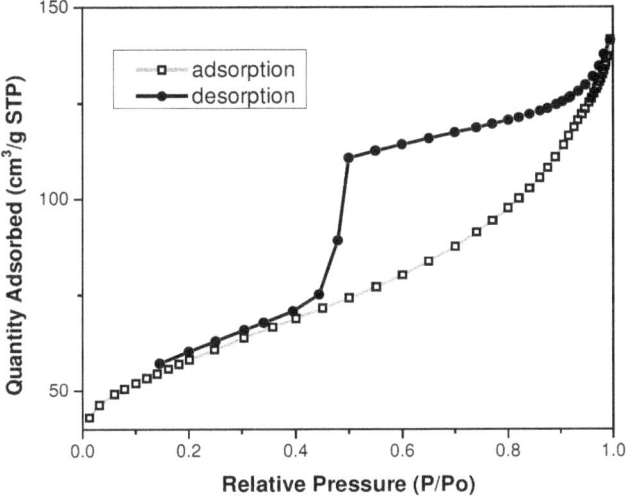

Figure 3. Nitrogen adsorption and desorption isotherm for PKS-Fe 1000 °C.

The nitrogen sorption isotherm for PKS-Fe 1000 showed a type IV isotherm, with a hysteresis loop associated with capillary condensation that took place in mesopores. The type IV isotherm with a hysteresis loop was initiated at a relative pressure, P/P_o, of approximately 0.17 and closing at 1.0. The hysteresis depicted the H2 class of hysteresis loops, which indicated a constriction associated with disordered carbon [62]. This provides a clear understanding of the types of pores that the graphitized carbon was evolved into as H2 hysteresis loops occurred when the pores were narrow mouth shapes (ink-bottles similar to pores), which resulted in a delay during desorption. This type of isotherm agrees with the data reported by Thompson et al. (2015), indicating the typical biomass responsible for mesopore structures [25].

Further insight into the detailed microstructures of the samples was elucidated with high-resolution transmission electron microscope (HRTEM) image analysis. Figure 4a shows an image of the PKS Fe-1000 sample; a visible set of core lattice fringes arising from graphitic carbon was observed [28,63]. This demonstrates a high degree of crystallinity in the PKS Fe-1000. On the other hand, the PKS control samples exhibited disordered microstructures typical of amorphous carbon, as shown in Figure 4b. Figure 4c shows the distinct lattice distance of graphite (002) with d-spacing of 0.33 nm, which implied the presence of a well-graphitized structure that is beneficial for electron transport [63]. The estimated d-spacing from HRTEM is in good agreement with the XRD data. In summary, HRTEM analysis confirmed the successful modification of lignocellulosic waste into graphite.

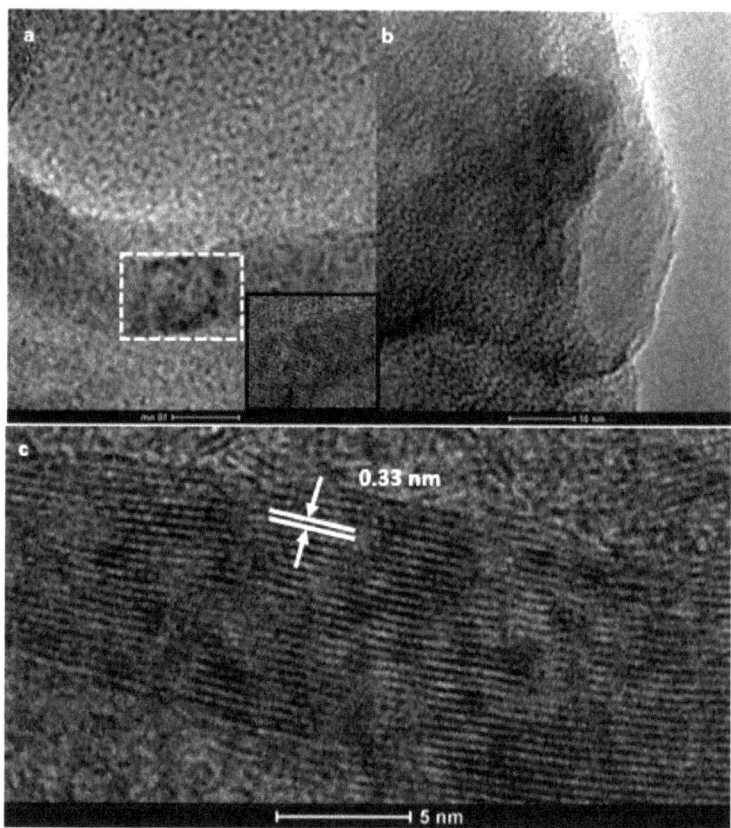

Figure 4. HRTEM image: (**a**) PKS Fe-1000 at 10 nm resolution; (**b**) PKS control-1000 at 10 nm resolution; (**c**) PKS Fe-1000 at 5 nm resolution.

The chemical composition of oil palm waste is a major factor that influences the degree of graphitization of carbon sources. Previous research work has suggested that selecting plant-based biomass with high lignin fraction, low cellulose fraction, low oxygen and high nitrogen content is important to ensure a high degree of graphitization [64].

PKS is made up of mainly cellulose, hemicellulose, Klason lignin, wax and ash. The composition in each component is different for each type of waste. A previous researcher conducted several tests to determine the composition of palm waste using a procedure recommended by the US National Renewable Energy Laboratory similar to ASTM E1758-01 [65]. It was found that PKS dry basis composition is made up of 14.7% cellulose, 16.4% hemicellulose, 53.6% Klason lignin, 2.3% wax, 2.3% ash and 10.7% (wt.%, dry basis) other components [65]. Meanwhile, the ultimate analysis of PKS at 380 °C (wt.%, daf at 380 °C) shows that it is made up of 80.9% carbon, 4.8% hydrogen, 0.7% nitrogen, 13.8% oxygen and 0.1% sulphur [64]. The data show that PKS contains a high amount of lignin and carbon but a low amount of cellulose, comparable to the observations in other research work, suggesting the successful transformation of amorphous into a graphitic structure [64,65].

3.3. The Possible Mechanisms for Graphitization

Several mechanisms have been proposed for the transformation of amorphous carbon structures into crystalline graphite with the addition of a catalyst. The addition of a metal catalyst has a significant remarkable impact on lowering the graphitization temperature [66]. Upon heat treatment, carbon atoms react with the catalyst to form several carbides, the decomposition of which results in the formation of graphite crystals. This process moves, assembles and disintegrates carbon atoms into graphite [66,67]. Dissolution-precipitation mechanisms have been proposed for the formation of graphite from amorphous carbon [66,68]. At a certain temperature, disordered carbon tends to diffuse and dissolve into metal or metal carbide and saturated carbon solubility reached under an equilibrium situation. With decreasing temperature, the metal saturated with the disordered carbon is supersaturated with carbon. Consequently, carbon precipitates in the form of graphite crystal because graphite is a highly ordered carbon with the lowest Gibbs free energy [45,59].

4. Conclusions

In summary, a biographitic carbon material with high crystallinity and high S_{BET} was successfully synthesized using palm kernel shell waste. Palm kernel shell is currently being utilized as a carbon precursor and ferum nitrate as a graphitization catalyst. The degree of graphitization can be regulated by changing the temperature and raw material, and also the type of catalyst. A higher heat treatment temperature of 1000 °C is responsible for a higher degree of graphitization. XRD data reported $2\theta = 26°$ for all samples, but higher and sharper peaks were noticeable for the PKS prepared with an iron catalyst with a lower Id/Ig ratio, which indicated high crystallinity and low defects. The values of the d_{002} spacing of all the samples were between 0.3325 and 0.3351, close to the value, 3.354, of pure graphite and less than 3.44 for disordered carbon, proving that the production of graphite from lignocellulosic material is a promising alternative for reusing waste material, for a circular economy and zero-waste focal point.

Author Contributions: Conceptualization, N.H.J. and R.O.; methodology, A.S.K.; software, A.S.K.; validation, R.O. and N.H.J.; formal analysis, A.S.K.; investigation, A.S.K. and R.O.; resources, N.H.J.; data curation, A.S.K.; writing—original draft preparation, A.S.K.; writing—review and editing, R.O. and N.H.J.; supervision, R.O.; project administration, N.H.J.; funding acquisition, N.H.J. All authors have read and agreed to the published version of the manuscript.

Funding: The authors acknowledge the funding support from the FRGS Ministry of Higher Education grant (Ref: FRGS/1/2015/TK05/UNIKL/02/1).

Institutional Review Board Statement: Not applicable.

Informed Consent Statement: Not applicable.

Data Availability Statement: Data are available on request from the corresponding authors. The data are not publicly available due to the need for further research work.

Acknowledgments: The authors are grateful for the financial support from the FRGS grant and Bell KSL Sawit. Sdn. Bhd for supplying the raw material for this project.

Conflicts of Interest: The authors declare no conflict of interest. The funders had no role in the design of the study; in the collection, analyses, or interpretation of the data; in the writing of the manuscript; or in the decision to publish the results.

References

1. Ma, Z.; Yang, Y.; Ma, Q.; Zhou, H.; Luo, X.; Liu, X.; Wang, S. Evolution of the chemical composition, functional group, pore structure and crystallographic structure of bio-char from palm kernel shell pyrolysis under different temperatures. *J. Anal. Appl. Pyrolysis* **2017**, *127*, 350–359. [CrossRef]
2. Dungani, R.; Aditiawati, P.; Aprilia, S.; Yuniarti, K.; Karliati, T.; Suwandhi, I.; Sumardi, I. Biomaterial from oil palm waste: Properties, characterization and applications. *Palm Oil* **2018**. [CrossRef]
3. Book&Claim. *Green Palm Sustainability*; Book&Claim Ltd.: Hull, UK, 2016; Available online: https://greenpalm.org/about-palm-oil/what-is-palm-oil/what-is-palm-oil-used-for (accessed on 1 October 2020).
4. Hamzah, N.; Tokimatsu, K.; Yoshikawa, K. Solid fuel from oil palm biomass residues and municipal solid waste by hydrothermal treatment for electrical power generation in Malaysia: A review. *Sustainability* **2019**, *11*, 1060. [CrossRef]
5. Dalton, S.; Mohamed, A.F.; Chikere, A.O. Status Evaluation of Palm Oil Waste Management Sustainability in Malaysia. *OIDA Int. J. Sustain. Dev.* **2017**, *10*, 41–48.
6. Pires, A.; Martinho, G. Waste hierarchy index for circular economy in waste management. *Waste Manag.* **2019**, *95*, 298–305. [CrossRef]
7. Hassan, M.A.; Abd-Aziz, S. *Waste and Environmental Management in the Malaysian Palm Oil Industry*; AOCS Press: Champaign, IL, USA, 2012; Volume 23. [CrossRef]
8. Choy, Y.K. Can palm oil waste be a solution to fossil fuel scarcity and environmental sustainability? A Malaysian case study provides the answer. *Waste Manag. Environ. VIII* **2016**, *1*, 97–108. [CrossRef]
9. Ślusarczyk, B.; Baryń, M.; Kot, S. Tire industry products as an alternative fuel. *Polish J. Environ. Stud.* **2016**, *25*, 1263–1270. [CrossRef]
10. Ungureanu, M.; Jozsef, J.; Brezoczki, V.M.; Monka, P.; Ungureanu, N.S. Research regarding the energy recovery from municipal solid waste in maramures county using incineration. *Processes* **2021**, *9*, 514. [CrossRef]
11. Bilan, Y.; Hussain, H.I.; Haseeb, M.; Kot, S. Sustainability and economic performance: Role of organizational learning and innovation. *Eng. Econ.* **2020**, *31*, 93–103. [CrossRef]
12. Kamal, S.; Jabarullah, N.H.; Othman, R. Catalytic graphitization of Oil Palm Frond using iron and silica. *Mater. Today Proc.* **2020**, *31*, 211–216. [CrossRef]
13. Chen, L.; Ji, T.; Mu, L.; Shi, Y.; Brisbin, L.; Guo, Z.; Khan, M.A.; David, P.; Zhu, J. Facile synthesis of mesoporous carbon nanocomposites from natural biomass for efficient dye adsorption and selective heavy metal removal. *RSC Adv.* **2016**. [CrossRef]
14. Ramachandran, V.; Ismail, F.S.; Noor, M.J.M.M.; Akhir, F.N.M.; Othman, N.; Zakaria, Z.; Hara, H. Extraction and intensive conversion of lignocellulose from oil palm solid waste into lignin monomer by the combination of hydrothermal pretreatment and biological treatment. *Bioresour. Technol. Rep.* **2020**, *11*, 100456. [CrossRef]
15. Bień, J. Production and use of waste-derived fuels in Poland: Current status and perspectives. *Prod. Eng. Arch.* **2021**, *27*, 36–41. [CrossRef]
16. Sinaga, O.; Saudi, M.H.M.; Roespinoedji, D.; Jabarullah, N.H. Environmental impact of biomass energy consumption on sustainable development: Evidence from ARDL bound testing approach. *Ekoloji* **2019**, *28*, 443–452.
17. Jabarullah, N.H. The controversy of biofuel versus fossil fuel. *Int. J. Adv. Appl. Sci.* **2016**, *3*, 11–14.
18. Tang, W.; Zhang, Y.; Zhong, Y.; Shen, T.; Wang, X.; Xia, X.; Tu, J. Natural biomass-derived carbons for electrochemical energy storage. *Mater. Res. Bull.* **2017**, *88*, 234–241. [CrossRef]
19. Xiao, P.-W.; Meng, Q.; Zhao, L.; Li, J.-J.; Wei, Z.; Han, B.-H. Biomass-derived flexible porous carbon materials and their applications in supercapacitor and gas adsorption. *Mater. Des.* **2017**, *129*, 164–172. [CrossRef]
20. Xi, Y.; Wang, Y.; Yang, D.; Zhang, Z.; Liu, W.; Li, Q.; Qiu, X. K_2CO_3 activation enhancing the graphitization of porous lignin carbon derived from enzymatic hydrolysis lignin for high performance lithium-ion storage. *J. Alloy. Compd.* **2019**, *785*, 706–714. [CrossRef]
21. Han, J.; Kwon, J.H.; Lee, J.-W.; Lee, J.H.; Roh, K.C. An effective approach to preparing partially graphitic activated carbon derived from structurally separated pitch pine biomass. *Carbon* **2017**, *118*, 431–437. [CrossRef]
22. Hou, L.; Hu, Z.; Wang, X.; Qiang, L.; Zhou, Y.; Lv, L.; Li, S. Hierarchically porous and heteroatom self-doped graphitic biomass carbon for supercapacitors. *J. Colloid Interface Sci.* **2019**, *540*, 88–96. [CrossRef]
23. Paun, V.A.; Popa, M.; Desbrieres, J.; Dragan, S.V.; Zegan, G.; Cioca, G. Liposome loaded chitosan hydrogels, a promising way to reduce the burst effect in drug release a comparativ analysis. *Mater. Plast.* **2016**, *53*, 590–593.
24. Xia, J.; Zhang, N.; Chong, S.; Li, D.; Chen, Y.; Sun, C. Three-dimensional porous graphene-like sheets synthesized from biocarbon via low-temperature graphitization for a supercapacitor. *Green Chem.* **2017**, *20*, 694–700. [CrossRef]
25. Thompson, E.G.; Danks, A.E.; Bourgeois, L.; Schnepp, Z. Iron-catalyzed graphitization of biomass. *Green Chem.* **2014**, *17*, 551–556. [CrossRef]
26. Yahayu, M.; Abas, F.Z.; Zulkifli, S.E.; Ani, F.N. Utilization of oil palm fiber and palm kernel shell in various applications. In *Sustainable Technologies for the Management of Agricultural Wastes*; Springer: Singapore, 2018; pp. 45–56.
27. Zhang, C.; Lu, G.; Sun, Z.; Yu, J. Catalytic graphitization of carbon/carbon composites by lanthanum oxide. *J. Rare Earths* **2012**, *30*, 128–132. [CrossRef]
28. Jiang, F.; Yao, Y.; Natarajan, B.; Yang, C.; Gao, T.; Xie, H.; Wang, Y.; Xu, L.; Chen, Y.; Gilman, J.; et al. Ultrahigh-temperature conversion of biomass to highly conductive graphitic carbon. *Carbon* **2019**, *144*, 241–248. [CrossRef]

29. Kalyoncu, R.S. Graphite. *U.S. Geol. Surv. Miner. Yearb. Met. Miner.* **2000**, *1*, 1076.
30. Kim, T.; Lee, J.; Lee, K.-H. Full graphitization of amorphous carbon by microwave heating. *RSC Adv.* **2016**, *6*, 24667–24674. [CrossRef]
31. Um, J.H.; Ahn, C.-Y.; Kim, J.; Jeong, M.; Sung, Y.-E.; Cho, Y.-H.; Kim, S.-S.; Yoon, W.-S. From grass to battery anode: Agricultural biomass hemp-derived carbon for lithium storage. *RSC Adv.* **2018**, *8*, 32231–32240. [CrossRef]
32. Shi, J.; Wang, Y.; Du, W.; Hou, Z. Synthesis of graphene encapsulated Fe3C in carbon nanotubes from biomass and its catalysis application. *Carbon* **2016**, *99*, 330–337. [CrossRef]
33. Thambiliyagodage, C.; Ulrich, S.; Araujo, P.; Bakker, M.G. Catalytic graphitization in nanocast carbon monoliths by iron, cobalt and nickel nanoparticles. *Carbon* **2018**, *134*, 452–463. [CrossRef]
34. Qin, H. Catalytic Graphitization Strategy for the Synthesis of Graphitic Carbon Nanocages and Electrochemical Performance. *Int. J. Electrochem. Sci.* **2017**, *12*, 10599–10604. [CrossRef]
35. Käärik, M.; Arulepp, M.; Karelson, M.; Leis, J. The effect of graphitization catalyst on the structure and porosity of SiC derived carbons. *Carbon* **2008**, *46*, 1579–1587. [CrossRef]
36. Shi, W.; Chang, B.; Yin, H.; Zhang, S.; Yang, B.; Dong, X. Crab shell-derived honeycomb-like graphitized hierarchically porous carbons for satisfactory rate performance of all-solid-state supercapacitors. *Sustain. Energy Fuels* **2019**, *3*, 1201–1214. [CrossRef]
37. Gutiérrez-Pardo, A.; Ramírez-Rico, J.; Cabezas-Rodríguez, R.; Martínez-Fernández, J. Effect of catalytic graphitization on the electrochemical behavior of wood derived carbons for use in supercapacitors. *J. Power Sources* **2015**, *278*, 18–26. [CrossRef]
38. Khokhlova, G.P.; Barnakov, C.N.; Malysheva, V.Y.; Popova, A.N.; Ismagilov, Z.R. Effect of heat treatment conditions on the catalytic graphitization of coal-tar pitch. *Solid Fuel Chem.* **2015**, *49*, 66–72. [CrossRef]
39. McKee, D.W. Carbon and Graphite Science. *Annu. Rev. Mater. Res.* **1973**, *3*, 195–231. [CrossRef]
40. Liu, Y.; Liu, Q.; Gu, J.; Kang, D.; Zhou, F.; Zhang, W.; Wu, Y.; Zhang, D. Highly porous graphitic materials prepared by catalytic graphitization. *Carbon* **2013**, *64*, 132–140. [CrossRef]
41. Sevilla, M.; Fuertes, A. Graphitic carbon nanostructures from cellulose. *Chem. Phys. Lett.* **2010**, *490*, 63–68. [CrossRef]
42. Nettelroth, D.; Schwarz, H.-C.; Burblies, N.; Guschanski, N.; Behrens, P. Catalytic graphitization of ordered mesoporous carbon CMK-3 with iron oxide catalysts: Evaluation of different synthesis pathways. *Phys. Status Solidi* **2016**, *213*, 1395–1402. [CrossRef]
43. Hoekstra, J.; Beale, A.; Soulimani, F.; Versluijs-Helder, M.; Geus, J.W.; Jenneskens, L.W. Base Metal Catalyzed Graphitization of Cellulose: A Combined Raman Spectroscopy, Temperature-Dependent X-ray Diffraction and High-Resolution Transmission Electron Microscopy Study. *J. Phys. Chem. C* **2015**, *119*, 10653–10661. [CrossRef]
44. Hoekstra, J.; Beale, A.; Soulimani, F.; Versluijs-Helder, M.; van de Kleut, D.; Koelewijn, J.M.; Geus, J.W.; Jenneskens, L.W. The effect of iron catalyzed graphitization on the textural properties of carbonized cellulose: Magnetically separable graphitic carbon bodies for catalysis and remediation. *Carbon* **2016**, *107*, 248–260. [CrossRef]
45. Yan, Q.; Li, J.; Zhang, X.; Hassan, E.B.; Wang, C.; Zhang, J.; Cai, Z. Catalytic graphitization of kraft lignin to graphene-based structures with four different transitional metals. *J. Nanoparticle Res.* **2018**, *20*, 223. [CrossRef]
46. Rashidi, N.A.; Yusup, S. Co-valorization of delayed petroleum coke—palm kernel shell for activated carbon production. *J. Hazard. Mater.* **2021**, *403*, 123876. [CrossRef]
47. Chin, C.O.; Yang, X.; Paul, S.C.; Susilawati; Wong, L.S.; Kong, S.Y. Development of thermal energy storage lightweight concrete using paraffin-oil palm kernel shell-activated carbon composite. *J. Clean. Prod.* **2020**, *261*, 121227. [CrossRef]
48. Yang, X.; Li, F.; Xia, M.; Luo, F.; Jiang, Y. Investigation on the micro-structure and adsorption capacity of cellulosic biomass carbon based montmorillonite composite. *Microporous Mesoporous Mater.* **2018**, *256*, 18–24. [CrossRef]
49. Pudukudy, M.; Yaakob, Z.; Takriff, M.S. Methane decomposition over unsupported mesoporous nickel ferrites: Effect of reaction temperature on the catalytic activity and properties of the produced nanocarbon. *RSC Adv.* **2016**, *6*, 68081–68091. [CrossRef]
50. Sevilla, M.; Sanchís, C.; Solis, M.S.; Morallón, E.; Fuertes, A.B. Synthesis of Graphitic Carbon Nanostructures from Sawdust and Their Application as Electrocatalyst Supports. *J. Phys. Chem. C* **2007**, *111*, 9749–9756. [CrossRef]
51. Sevilla, M.; Lecea, C.S.M. D.; Solis, M.S.; Morallón, E.; Fuertes, A.B. Solid-phase synthesis of graphitic carbon nanostructures from iron and cobalt gluconates and their utilization as electrocatalyst supports. *Phys. Chem. Chem. Phys.* **2008**, *10*, 1433–1442. [CrossRef] [PubMed]
52. Johnson, M.; Faber, K. Catalytic graphitization of three-dimensional wood-derived porous scaffolds. *J. Mater. Res.* **2011**, *26*, 18–25. [CrossRef]
53. Liu, W.-J.; Tian, K.; He, Y.-R.; Jiang, H.; Yu, H.-Q. High-Yield Harvest of Nanofibers/Mesoporous Carbon Composite by Pyrolysis of Waste Biomass and Its Application for High Durability Electrochemical Energy Storage. *Environ. Sci. Technol.* **2014**, *48*, 13951–13959. [CrossRef]
54. Zhao, J.; Liu, Y.; Quan, X.; Chen, S.; Yu, H.; Zhao, H. Nitrogen-doped carbon with a high degree of graphitization derived from biomass as high-performance electrocatalyst for oxygen reduction reaction. *Appl. Surf. Sci.* **2017**, *396*, 986–993. [CrossRef]
55. Yang, J.; Zuo, S. Facile synthesis of graphitic mesoporous carbon materials from sucrose. *Diam. Relat. Mater.* **2019**, *95*, 1–4. [CrossRef]
56. Qiu, T.; Yang, J.-G.; Bai, X.-J.; Wang, Y.-L. The preparation of synthetic graphite materials with hierarchical pores from lignite by one-step impregnation and their characterization as dye absorbents. *RSC Adv.* **2019**, *9*, 12737–12746. [CrossRef]
57. Vázquez-Santos, M.B.; Geissler, E.; László, K.; Rouzaud, J.-N.; Martínez-Alonso, A.; Tascon, J.M.D. Comparative XRD, Raman, and TEM Study on Graphitization of PBO-Derived Carbon Fibers. *J. Phys. Chem. C* **2012**, *116*, 257–268. [CrossRef]

58. Zhou, X.; Chen, F.; Bai, T.; Long, B.; Liao, Q.; Ren, Y.; Yang, J. Interconnected highly graphitic carbon nanosheets derived from wheat stalk as high performance anode materials for lithium ion batteries. *Green Chem.* **2016**, *18*, 2078–2088. [CrossRef]
59. Chen, C.; Sun, K.; Wang, A.; Wang, S.; Jiang, J. Catalytic Graphitization of Cellulose Using Nickel as Catalyst. *BioResources* **2018**, *13*, 3165–3176. [CrossRef]
60. Lim, Y.; Chu, J.H.; Lee, D.H.; Kwon, S.-Y.; Shin, H. Increase in graphitization and electrical conductivity of glassy carbon nanowires by rapid thermal annealing. *J. Alloy. Compd.* **2017**, *702*, 465–471. [CrossRef]
61. Major, I.; Pin, J.-M.; Behazin, E.; Rodriguez-Uribe, A.; Misra, M.; Mohanty, A. Graphitization of Miscanthus grass biocarbon enhanced by in situ generated FeCo nanoparticles. *Green Chem.* **2018**, *20*, 2269–2278. [CrossRef]
62. Zhang, P. Adsorption and Desorption Isotherms. 2016. Available online: http://www.keresearchgroup.com/uploads/4/8/4/5/48456521/160903_introduction_to_bet_isotherms.pdf (accessed on 15 February 2021).
63. Wu, F.; Huang, R.; Mu, D.; Wu, B.; Chen, Y. Controlled synthesis of graphitic carbon-encapsulated α-Fe_2O_3 nanocomposite via low-temperature catalytic graphitization of biomass and its lithium storage property. *Electrochim. Acta* **2016**, *187*, 508–516. [CrossRef]
64. Liu, Y.; Nie, Y.; Lu, X.; Zhang, X.; He, H.; Pan, F.; Zhou, L.; Liu, X.; Ji, X.; Zhang, S. Cascade utilization of lignocellulosic biomass to high-value products. *Green Chem.* **2019**, *21*, 3499–3535. [CrossRef]
65. Yuliansyah, A.T.; Hirajima, T. Efficacy of hydrothermal treatment for production of solid fuel from oil palm wastes. *Resour. Manag. Sustain. Agric.* **2012**. [CrossRef]
66. Liu, D.; Zhao, X.; Su, R.; Hao, Z.; Jia, B.; Li, S.; Dong, L. Highly porous graphitic activated carbons from lignite via microwave pretreatment and iron-catalyzed graphitization at low-temperature for supercapacitor electrode materials. *Process.* **2019**, *7*, 300. [CrossRef]
67. Yu, Z.-L.; Xin, S.; You, Y.; Yu, L.; Lin, Y.; Xu, D.-W.; Qiao, C.; Huang, Z.-H.; Yang, N.; Yu, S.-H.; et al. Ion-catalyzed synthesis of microporous hard carbon embedded with expanded nanographite for enhanced lithium/sodium storage. *J. Am. Chem. Soc.* **2016**, *138*, 14915–14922. [CrossRef] [PubMed]
68. Kamal, A.S.; Othman, R.; Jabarullah, N.H. Preparation and synthesis of synthetic graphite from biomass waste: A review. *Syst. Rev. Pharm.* **2020**, *11*, 881–894.

Mining Industry Impact on Environmental Sustainability, Economic Growth, Social Interaction, and Public Health: An Application of Semi-Quantitative Mathematical Approach

Muhammad Mohsin [1], Qiang Zhu [1], Sobia Naseem [2,*], Muddassar Sarfraz [3,4,*] and Larisa Ivascu [4,5]

[1] School of Business, Hunan University of Humanities, Science and Technology, Loudi 417000, China; mohsinlatifntu@gmail.com (M.M.); zhuqiang1968@126.com (Q.Z.)
[2] School of Economics and Management, Shijiazhuang Tiedao University, Shijiazhuang 050043, China
[3] College of International Students, Wuxi University, Wuxi 214105, China
[4] Research Center for Engineering and Management, Politehnica University of Timisoara, 300191 Timisoara, Romania; larisa.ivascu@upt.ro
[5] Faculty of Management in Production and Transportation, Politehnica University of Timisoara, 300191 Timisoara, Romania
* Correspondence: Sobiasalamat@stdu.edu.cn (S.N.); muddassar.sarfraz@gmail.com (M.S.); Tel.: +86-18751861057 (M.S.)

Abstract: The mining industry plays a significant role in economic growth and development. Coal is a viable renewable energy source with 185.175 billion deposits in Thar, which has not been deeply explored. Although coal is an energy source and contributes to economic development, it puts pressure on environmental sustainability. The current study investigates Sindh Engro coal mining's impact on environmental sustainability and human needs and interest. The Folchi and Phillips Environmental Sustainability Mathematics models are employed to measure environmental sustainability. The research findings demonstrated that Sindh Engro coal mining is potentially unsustainable for the environment. The toxic gases (methane, carbon dioxide, sulfur, etc.) are released during operational activities. The four significant environment spheres (atmosphere, hydrosphere, biosphere, and lithosphere) are negatively influenced by Thar coal mining. The second part of the analysis results shows that human needs and interests have a positive and significant relationship except for human health and safety with Sindh Engro coal mining. Environmental pollution can be controlled by utilizing environmentally friendly coal mining operations and technologies. Plantation and ecological normalization can protect the species, flora, and fauna of the Thar Desert. The government of Pakistan and the provincial government of Sind should strictly check the adaptation of environmental standards. Furthermore, the researchers should explore the environmental issues and solutions so that coal mining becomes a cost-efficient and environmental-friendly energy source in Pakistan.

Keywords: sindh engro coal mine; environmental sustainability; health and safety; social interaction and quality of life; economic development

1. Introduction

The rapid increase of coal exploitation and utilization declares that coal is a significant energy source. The contribution of coal energy to an individual country's economy cannot be denied, but coal mining distracts environmental sustainability. The regional and global coal mining areas are being influenced by incomplete combustion of carbonaceous fuels, i.e., Black Carbon. Black carbon is a solar radiation absorber and heats the earth–atmosphere system. Methane (CH_4) is abundantly released during the coal-mining process, and its leakage due to imperfect blockage methods, cracks, and vents in mines also facilitate to continue leaking outside. According to Environmental Impact Assessment (EIA), methane gas emissions during mining are 1% of total greenhouse gas emissions globally. Greenhouse

(GHG) gas emission is a continuing problem in mining. According to Ohara, the Asian region increasingly contributes to GHG emissions [1]. The Regional Emission Inventory in Asia (REAS) was the first historical integration of present and future emissions in Asia by utilizing consistent techniques and methods. China inventory and emissions contribution are analyzed under three emission scenarios: Policy Success Case (PSC), Reference Case (REF), and Policy Failure Case (PFC). The total Asian emissions are affected by the contribution of China.

In China, the rapid emission is increased mainly due to coal combustion in the power plants and industrial sector. China mainly focused on production quantity while compromising on environmental sustainability. China will be able to decrease black carbon emission from 1.33 to 1.16 Tg [2]. Pakistan is going to compromise on environmental sustainability because of the electricity crisis. The coal production and consumption boost other economic sectors/industries, i.e., power plants, cement industry, coke use, and brick kilns. Most industries utilized coal as a substitute for petroleum products, which became cost-efficient and profitable for sectors. The rapid growth in demand for energy for sectorial and domestic use encourage the exploration and utilization of the energy deposits. According to vision 2025, with Chinese companies' collaboration, Pakistan will be self-sufficient in the power sector, electricity demand and supply, coal efficiency, and power generation. The coal consumption is increased annually in Pakistan as 32.91% in 2015, 3.57% in 2016, 23.92% in 2017, and 60.75% in 2018. Pakistan made a joint agreement with the China Pakistan Economic Corridor (CPEC) to minimize CO_2 emission and clean coal technology to maintain the sustainability of the environment [3–5].

Almost all counties with ongoing coal mining processes are expected to have environmental sustainability issues and unnecessary emissions. The territorial analysis of Silesian counties based on the Prevalent Vulnerability Index scoring also confirmed a negative relationship between coal mining and ecological sustainability [6]. The special report of the Intergovernmental Panel on Climate Change (IPCC) 2018 on the impact of global warming stated that environmental health at 1.5 °C above the preindustrial level could not be maintained. It has shown an underlined global intent and effort to reduce the global greenhouse emission by at least 45% until 2030 and brought it to be net-zero until 2050 [7,8]. Burke and Fishel [8] proposed the Coal Elimination Treaty (CET) to overcome greenhouse gas emissions and attain average global heating at 1.5 °C before 2050. In his research, three negotiation pathways are presented with the inclusion of the inspirational normative model 2017 treaty on the prohibition of nuclear weapons. The primary focus is the progressive stigmatization, prohibition, and coal elimination to avoid an unmanageable climatic future.

Coal mining positively influenced Pakistan's economic development by contributing 13% of commercial energy consumption. Coal is an inexpensive energy source with 185.175 billion tons of deposits more than petroleum, oil, LPG, and LNG. The enhancement of energy and Pakistan's economy is continuously increasing by coal deposit utilization over the last ten years under the Alternative Energy Development Plan (AEDP), China Pakistan Economic Corridor (CPEC), Vision 2025, Vision 2035, and INDC development projects (Masih, 2018). Pakistan and China are collaboratively working on the coalfields, and Pakistan will install 8.4GW (capacity) coalfield power plants (CFPP) until the end of 2021. The expected contribution in GHG emissions by CFPP plants is 67% or 63.6 MT $CO_2 a^{-1}$. The scarcity of electricity generation sources in Pakistan compels burning coal to meet the needs of the population. These coalfield explorations and burning coal will increase the electricity supply from 0.1% to 16.5% in Pakistan until 2050 [9,10]. However, Pakistan is facing significant challenges in sustainable waste management in coal mining. The disposal methods of coal mining are not up to the standards of any developed country. This unsustainable waste management is seriously contributing to environmental degradation globally and domestically. Global GHG emissions must be substantially lowered for a sustainable environment.

The Sindh Engro coalfield is the largest coalfield of Pakistan and considered one of the largest coal deposits (1.33 billion tonnes of coking coal) in a world that covers an area of 9000 KM2 approximately. As per the declaration of diagnostic macerals and petrographic, Thar coal is humic and predominantly features huminite with significant amounts of liptinite and low quantities of inertinite macerals. The Thar coal represents predominantly topogenous mires deposited under anaerobic conditions with limited thermal and oxidative tissue destruction. Coal mining near the residential area of Sindh has a prodigious impact on the environment, ecological habitats, society, and the country's economy. The global environment integrates the global living beings with their relationship with the atmosphere, hydrosphere, biosphere, and lithosphere. These four components of the environment enable us to cover all environmental sustainability areas. The open-pit coal mining's remediation of destruction has been the focus of several studies. Previous literature has considered the coalfields as an integral part of the global economy. Still, the greenhouse gas emission from coal mining has exceeded the remaining sectors of the global carbon budget [2,9,11–13].

The environmental pollution level is increasing in Pakistan day by day. We keenly explored the particular directions of coal mining to get the facts about pollution hub to control the pollution as much as possible by using green ways of production. The research aims to explore the coal mining contribution to environmental pollution comprehensively. Pakistan has several coal ranges, but in this research, we used the Sindh Engro coal mine as a study sample. There are three fundamental reasons to select the Sindh Engro coalfield for this research: firstly, the Sindh Engro coalfield is the largest coal deposit area in Pakistan. Secondly, these coal deposits are mined with the collaboration of foreign companies as China. As per foreign interest, this coal range uses developed technology compared to other coal ranges. A developed technology user area suffers environmental issues, and the undeveloped or under-developed mining methods have more chance. As per the large deposit of coal in the Sindh coalfield, there is more chance of uptrend environmental pollution. If the facts and figures confirm the positive relationship between environmental pollution and coal mining, environmentalists can take practical steps to control it. The validity of this research and control methods will be checked quickly and applied to other coal ranges for attaining a better quality of the environment. The existing literature is generally based on air pollution, water pollution, and CO_2 emission. Coal mining influences all spheres of the environment (atmosphere, biosphere, hydrosphere, and lithosphere), which we will explore individually in this research. The range of research from the perspective of the environment influencing factors is comparatively wider than the existing coal mining literature.

2. Study Background
2.1. Coal Mining and Atmosphere

The earth's atmosphere is barricaded by the layer of gases known as air. The atmosphere retained the gravity of the earth and formatting the planetary surroundings. The specific pressure of air for liquid water on the earth's surface, absorption of toxic and ultraviolet radiations, heat retention, and normalization of diurnal temperature variation are controlled by the atmosphere to protect earth habitants. The suspended particular matters (SPM) and respirable particulate matter (RPM) of air pollution are observed in the air of coal mine surroundings as well as projected areas [14]. The air contains different gases at a specific level for healthy atmosphere equilibrium. In this era of development, the atmosphere's equilibrium has changed, and humans also contributed to this change by industrialization, mining, and environmentally opposed technology. The coal mining process and burning coal emit particulates, arsenic, and carbon dioxide [15,16]. The environment influences underground mining by reducing methane (CH_4), which is growing faster than carbon dioxide. The deeper coal mines contain a higher amount of methane (CH_4) deposits. During mining operational activities, methane gas is released into the air and becomes part of the atmosphere. The gas emission in operation and combustion coal become the reason for pillar collapse and roof falling. Mining operational and structural flaws also generate acidic mine drainage. The mine pillars cover large surface areas and the slow oxidation of a sulfur compound by exposure to air. The dumped sludge and wastes of the coal mining operation also increase the toxic effect under the surface. The sulfur oxidation compounds leaching of the resultant acid in surface dumps, creating a risk of rainwater percolating. The pyrite reaction with air and water with sulfuric acid then generates Acid Mine Discharge (AMD). The analytical part of this study is based on two parts: qualitative part (Folchi method) and quantitative part (Phillips Environmental Sustainability Mathematics (PESM) Model). Due to a combination of qualitative and quantitative features, this research confirmed its semi-quantitative quality. The Folchi method covered almost all environment influenced factors as per four significant spheres of environment, i.e., atmosphere, biosphere, hydrosphere, and lithosphere. Table 1 presented all possible coal mining environments and impacting factors specification. Table 2 contains the numbering according to the intensity of influencing factors for the indexing process. This table also explains the measuring scale of the individual factor for the convenience of readers. Table 3 is most important to understand for interlinking before and after processes. Table 3 is technically maintained with possible maximum, medium, and minimum points by Folchi. The values from Table 2 as per the contribution scale of environmental factors are multiplied with Table 3, and the result of the multiplication is Table 4. The aggregate scores of an individual factor in Table 4 declared the actual contribution of these factors in the environment for the Sindh Engro coalfield. The first qualitative part and indexing are completed. The second part, the Phillips Environmental Sustainability Mathematics (PESM) Model, is started. The expected scores and actual scores are put in the Phillips Environmental Sustainability Mathematics formula. The final figure decides the sustainable or unsustainable contribution of coal mining in environmental sustainability. As per our analysis, the Environmental contribution is less than Human needs and interests. The Sindh Engro coal mines confirmed an unsustainable relationship with the environment. The Sindh Engro coal mines have the largest deposit of coal and explore coal with foreign collaboration. Due to foreign association to digging coal, this coalfield utilized well-developed technology to explore the coal deposits. The unsustainable environmental contribution can ease the way to explore less developed coal mines and eradicate the emissions-friendly resources that will enable the sustainable contribution in the environment and economy.

The dissolution of acids and heavy metals, i.e., copper, lead, and mercury, in-ground or surface water, have an acidic reaction with soil.

Table 1. Mining environment categories and impact factors.

Sr. No.	Coal Mining Environment	Coal Mining Impacting Factors
1	Health and safety of human	Land dispossession and potential resources
2	Social interaction and quality of life	Exposition, visibility of the pit
3	Water pollution	Above-ground water pollution/depletion
4	Air pollution	Underground water pollution/depletion
5	Soil Erosion	Increase in vehicular traffic
6	Biodiversity loss	Atmospheric release of gas and coal dust
7	Above-ground interferences	Mine tailing spills
8	Underground interferences	Level of noise
9	Aesthetic degradation	Vibration of ground
10	Noise pollution	The livelihood of the local workforce
11	Economy	Contribution in GDP

Table 2. The magnitude of impacting factors.

Impacting Factors	Scenario	Magnitude
Land dispossession and potential resources	Parks, protected areas	8–10
	Urban/rural area	6–8
	wildlife, agro-diversity	3–6
	Industrial area	1–3
Exposition, visibility of the pit	Can be seen from inhabited areas	6–10
	Can be seen from the main roads	2–6
	Not visible	1–2
Underground water pollution/depletion	Interference with lakes and rivers	6–10
	Interferences with the non-relevant water system	3–6
	No interference	1–3
Above-ground water pollution/depletion	Surface water pollution/Decreasing water (physicochemical, biological) quality	5–10
	Water table deep and permeable grounds	2–5
	Water table deep and unpermeable grounds	1–2
Increase in vehicular traffic	Increase of 200%	6–10
	Increase of 100%	3–6
	No interference	1–3
Atmospheric release of gas and coal dust	Free emissions in the atmosphere	7–10
	Emission around the given reference values	2–7
	Emission well below the given reference values	1–2
Mine tailings spills	Toxic gases releasing during process	
	No blast design and no clearance procedures	9–10
	Blast design and no clearance procedures	4–9
	Blast design and clearance procedures	1–4
Level of noise	<141 db.	8–10
	<131 db.	4–8
	<121 db.	1–4
Vibration of ground	Cosmetic damage, above threshold	7–10
	Tolerability threshold	3–7
	Values under tolerability threshold	1–3
The livelihood of local workforce	Job opportunities	
	High	7–10
	Medium	3–6
	Low	1–2
Contribution in GDP	Level of GDP contribution	
	High	7–10
	Medium	3–6
	Low	1–2

Table 3. The correlation matrix of impacting factors and component magnitude.

Impact Factors	Human Health and Safety	Social Interaction and Quality of Life	Water Pollution	Air Pollution	Soil Erosion	Biodiversity Loss	Above-Ground Interferences	Underground Interferences	Environmental Components	Noise Pollution	Economy
Land dispossession and potential resources	Med	Min	Nil	Nil	Max	Min	Nil	Nil	Aesthetic Degradation	Nil	Nil
	0.8	0.77	0	0	5.71	0.63	0	0	2.86	0	0
Exposition, visibility of the pit	Nil	Min	Nil	Nil	Med	Nil	Nil	Nil	Max	Min	Nil
	0	0.77	0	0	2.86	0	0	0	2.86	2	0
Above-ground water pollution/depletion	Max	Nil	Max	Nil	Nil	Max	Med	Nil	Max	Nil	Nil
	1.6	0	4.44	0	0	2.5	6.67	0	2.86	0	0
Underground water pollution/depletion	Min	Nil	Max	Nil	Nil	Nil	Nil	Med	Nil	Nil	Nil
	0.4	0	4.44	0	0	0	0	6.67	0	0	0
Increase in vehicular traffic	Max	Max	Nil	Nil	Min	Max	Nil	Nil	Min	Nil	Nil
	1.6	3.08	0	0	1.43	2.5	0	0	0.71	0	0
Atmospheric release of gas and coal dust	Max	Min	Min	Max	Nil	Max	Min	Nil	Min	Nil	Nil
	1.6	0.77	1.11	10	0	2.5	3.33	0	0.71	0	0
Mine tailing spills	Max	Nil	Nil	Nil	Nil	Med	Nil	Nil	Nil	Nil	Nil
	1.6	0	0	0	0	1.25	0	0	0	0	0
Level of noise	Med	Max	Nil	Nil	Nil	Min	Nil	Nil	Nil	Max	Nil
	0.8	3.08	0	0	0	0.63	0	0	0	8	0
Vibration of ground	Max	Med	Nil	Nil	Nil	Nil	Nil	Min	Nil	Nil	Nil
	1.6	1.54	0	0	0	0	0	3.33	0	0	0
The livelihood of the local workforce	Nil	Nil	Nil	Nil	Nil	Nil	Nil	Nil	Nil	Nil	Max
	0	0	0	0	0	0	0	0	0	0	6.67
Contribution in GDP	Nil	Nil	Nil	Nil	Nil	Nil	Nil	Nil	Nil	Nil	Min
	0	0	0	0	0	0	0	0	0	0	3.33
Total	10	10	10	10	10	10	10	10	10	10	10

Table 4. Final weighted impact scores of Sindh Engro coal mining.

	Environmental Components									
	Human Health and Safety	Social Interaction and Quality of Life	Water Pollution	Air Pollution	Soil Erosion	Biodiversity Loss	Above-Ground Interferences	Underground Interferences	Aesthetic Degradation	Noise Pollution	Economy
Model's Abbreviations	H_{NJ1}	H_{NJ2}	H	A_1	L_1	B_1	L_2	L_3	L_4	A_2	H_{NJ3}
Land dispossession and potential resources	6.4	6.16	0	0	45.68	5.04	0	0	22.88	0	0
Exposition, visibility of the pit	0	4.62	0	0	17.16	0	0	0	17.16	12	0
Above-ground water pollution/depletion	12.8	0	35.52	0	0	20	53.36	0	22.88	0	0
Underground water pollution/depletion	3.6	0	39.96	0	0	0	0	60.03	0	0	0
Increase in vehicular traffic	14.4	27.72	0	0	12.87	22.5	0	0	6.39	0	0
Atmospheric release of gas and coal dust	16	7.7	11.1	100	0	25	33.3	0	7.1	0	0
Mine tailing spills	14.4	0	0	0	0	11.25	0	0	0	0	0
Level of Noise	6.4	24.64	0	0	0	5.04	0	0	0	64	0
Vibration of ground	12.8	12.32	0	0	0	0	0	26.64	0	0	0
Livelihood of local workforce	0	0	0	0	0	0	0	0	0	0	33.35
Contribution in GDP	0	0	0	0	0	0	0	0	0	0	9.99
Total	86.8	83.16	86.58	100	75.71	88.83	86.66	86.67	76.41	76	43.34

2.2. Coal Mining Impact on Biosphere

The local biodiversity suffers tremendous pressure from mining activities. The coal mining activities endangered the habitats and necessary ecosystem. The persistence and bio-accumulative nature of living beings is extremely endangered by potentially toxic elements (PTEs) with their toxic contamination and pollution [17,18]. The potentially toxic elements (PTEs) threatened the natural ecosystem by weathering and eroding parent rocks and ore deposits while susceptible anthropology by waste or tailing deposits, agrochemical, and industrial effluents. The discharging processes of potentially toxic elements in the ecosystem are mobilized by physicochemical and microbial factors, passing through divergent environmental cubicles [19,20]. The Indigofera, Indigofera oblongifolia, Indigofera argentea, Aristida funiculate, Convolvulus portraits, Cassia Italica, Dipterygium glaucum, and Digera ardencies, while plant communities Aerva javanica, Calligonum polygon sides, and Leptadenia pyrotechnica species are endangered due to the intensive environment of Thar. Animal grazed these plants at the pre-and post-monsoon periods when the ephemerals were not available due to the drought period. To protect these species, environmental sustainability with appropriate environmental conditions is necessary. Shortage of water resources, grazing trees, and plants is also an alarming situation for 35 mammalian fauna of Thar Pakistan. According to a wildlife ecologist report, Indian wild ass (Equus hemionus), Striped hyaena (Hyaena hyaena), chinkara (Gazella bennettii), and nilgai or blue bull (Boselaphus tragocamelus) are near threatened. The species are endangered due to Thar coal power projects and coal mining operations [21–23].

2.3. Coal Mining Impact on Hydrosphere

Mine excavation possesses water influx by rainfall or interception of groundwater flows. The minor portion of this water is used for processing and dust suppression and remaining pumped out due to its unwanted mining feature. The unwanted water feature is contaminated by particulate matter, oil and grease, un-burnt explosives, and other chemicals. The high rate of pyrites in coal indicates the toxic and acidic water that can affect the water resources by discharging it into nearby streams [24]. Whether under or above the earth's surface, the combined mass of water is known as the hydrosphere. The hydrosphere layer is disturbing by polluted and environmentally unfriendly human beings' activities. Mining is one activity that will unstabilize the water quality and increase water pollution. The underground and above groundwater resources significantly affect the irrigation system due to metal concentration in crops and soils. The increasing metal association level, metal partitioning, soil-to-plant bio-concentration factor (BCF), and gas concentration caused severe health issues by vegetable ingestion [7,17]. Severe pollution is observed in nearby streams, water, soil, and the cultivated crops over the coal mining area. By employing organic geochemical and petrographic analyses on critical petrographic facies and diagnostic macerals, the coal seams in Thar coalfield, coal rank, and hydrocarbon generation potential belong to predominant type III kerogen (gas-prone) grading into mixed types II–III kerogens (oil and gas-prone). The microscopic observation declared organic matter in Thar coal, which is dominated by huminite (woody) with a significant amount of oil-prone liptinite macerals and low inertinite macerals. The Tissue Preservation Index (TPI) and high Gelification Index (GI) demonstrated the consistency of marsh-wet forest swamp freshwater environment and limnotelmatic conditions.

2.4. Coal Mining Impact on Lithosphere

The lithosphere of the earth is composed of the crust, and the upper mental portion of it behaves elastically on time scales. The surface's mineralogy and chemistry are distinguished between the crust and upper mental. The mines and minerals belong to the continental crust with layers of igneous, sedimentary, and metamorphic rocks [24,25]. Soil erosion, inner and outer interferences, and land/aesthetic degradation are part of it. Soil erosion and land degradation primarily lead the topsoil and organic matter loss. The topsoil and organic matter are fundamental ingredients of planting. Generally, water, wind, and mass movements play a significant role to move soil erosion. Climate change is a key factor of global soil erosion degradation. The suitable conditions of soil erosion are rainfall (frequency, duration, and intensity), winds (level of intensity, directions of wind and frequency), and water (underground and above ground). The Thar coal mining areas fulfill all the soil erosion conditions, which negatively affect environmental sustainability. Natural geotextile formulating by organic material is more effective and beneficial to reduce soil erosion in working stations. In the last two decades, soil erosion, soil degradation, soil loss, and soil pollution are increasing, and environmental pollution accelerates it. The mineral process wastages, industrial wastages, and the pollution created by human activities contributed to it [26,27].

2.5. Human Needs and Interest and Coal Mining
2.5.1. Health and Safety of Human

The coal cutting, excavation, and transportation in the coal-mining field are involved in fatalities, injuries, and disease risk. The underground mining occupation is considered the most dangerous job, which becomes the reason for respiratory diseases such as infected lungs, skin allergy, deafness, heart burning, and psychological stress among the workers [28–30]. The poisonous gas hydrogen sulfide emits during coal mining, and its longer-term exposure caused eye irritates, headache, fatigue, lungs, and respiratory tract. The operational coal dust with toxic gases and improper management of wastages creates permanent health hazards for workers and the residential community [31,32]. Paul and Maiti [20] further explained that minors' behavioral motivation also affected the rate of accidents/injuries. By utilizing behavioral safety analysis on two neighboring coal mines of India, it is observed that the accident group of workers was dissatisfied with their job, which negatively influenced and risk alarming. The job dissatisfaction urges the worker to take more risk on the working station and behave unsafely. Panhwar confirmed the risk of health and safety in coal mining using the convenience sampling method and a structured questionnaire [33]. Panhwar et al. [30] approached 97 mines workers for data collection via a structured questionnaire regarding the work environment, physical health of miners, and protection policies of mines. According to the last five years, there were 4 deaths by roof falling, 1 death by stone fall down from mine shaft, 3 fatalities due to suffocation and inhaling toxic gases, 121 injuries by different accidents, and 1 causality through rope haulage pulley in Lakhra Coal Mines. The maximum health and safety risk is triggering because of a lack of awareness about labor law and mining laws in developing countries. According to the compliance law of mining in Pakistan, appropriate working conditions is compulsory for every worker. The mining workers generally belong to backward areas and are not well-educated, so they are deprived of full awareness about health and safety. The workers' unawareness leads to continuous fatalities and safety issues. The mining workers do not adequately get safety training before joining the job. The rate of causalities and accidents is higher for inexperienced and untrained workers than experienced workers [12,32,33]. The contaminated water is used as drinking water at the working station due to potable fresh water's unavailability. The hygiene and bad odor problems also occur because the proper toilets are not available at operational places [34–36].

2.5.2. Social Interaction and Quality of Life

The major development projects with lower environmental risk and higher social/economic contribution are considered beneficial for people. Pakistan is immediately trying to overcome energy crises, and achieving this goal, coal energy is exploited. Thar coal is mainly used to minimize the pressure of energy crises, while this project contributes to the development of Thar by generating new job opportunities. However, some social issues transpired in relation to Thar coal during a survey of the ground realities, interaction with Sindh Engro Coal Mining Company (SECMC) officials, and concerning villagers. The major concerns of society or the residents of Thar are the dumping or disposal of mining wastages. The villagers protest against the Gorano pond/dam construction to collect water waste and mines effluent. The coal mining activities have started in some blocks, and some are still untouched. The Gordano pond/dam can receive block II wastages and effluents, so how will the other blocks' effluents be disposed of after actively working with them? The construction of ponds for every mining block can destroy the soil quality, water quality, and environmental sustainability of Thar. No doubt, Thar coal mining contributes to social development. Simultaneously, constructing and ponding Gorano dam with saline water creates problems for ecology, morphology, biodiversity, hydrology, soil composition, and environmental sustainability.

3. Materials and Methods

3.1. Site Description

Pakistan has started to dig up the world's largest coal deposit in the Thar Desert with Chinese companies' collaboration. The Thar Desert coal station is located in the Mithi region, Tharparkar District, Sindh province of Pakistan (see Figure 1). This research has used Sindh Engro Coal Mining, Thar Desert, Pakistan, as a sample. This is a joint venture between Sindh and Engro Corporation's government, while Chinese Machinery Engineering Corporation is also part of this coal-mining project. The Thar Desert's mining coal is low-grade, brown, dirty coal known as lignite coal. The coal mining area is based on a rural community of almost 12 villages. The detailed operational activity in Thar Desert coal mining and an ariel view of the EPTL power plant in Thar are presented in Figures 2 and 3.

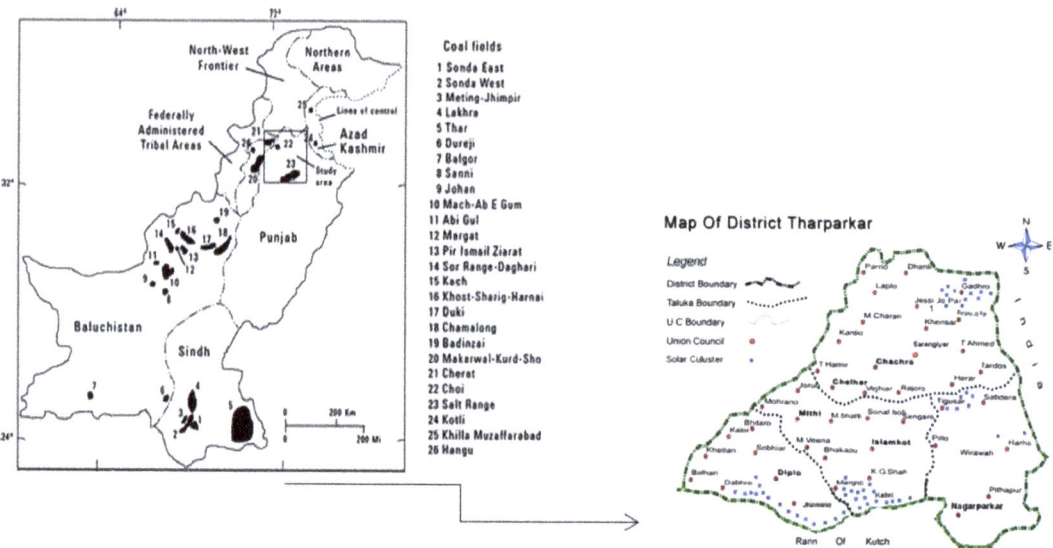

Figure 1. Coal fields in Pakistan and detailed map of district Tharparker.

Figure 2. An ariel view of EPTL power plant in Thar.

Figure 3. Operation at the Sindh Engro Coal Mining Company site in Thar Desert (source: Bloomberg).

3.2. Methodology

The current study research methodology is based on the Folchi method for environmental factorization and indexing, while the second is based on the Phillips Environmental Sustainability Mathematics (PESM) Model. In addition, the Folchi method has been used in PESM for data analysis.

3.2.1. Folchi Method

The Folchi method is technically designed for quantifying the impact of mining on the environment by operational activities (drilling, blasting, and collecting minerals). Previous studies confirm the accuracy of the analysis, but data analysis accuracy can be gained if the authorized personnel collect it [37,38].

The coal mining environment and factors affecting it are listed as presented in Table 1. The first column consists of coal mining environment categories, and the second column contains coal mining impacting factors. As per mining operation, all possible environmental factors are covered in it. In Figure 4, the Folchi method's step-by-step process is presented for a quick process visual.

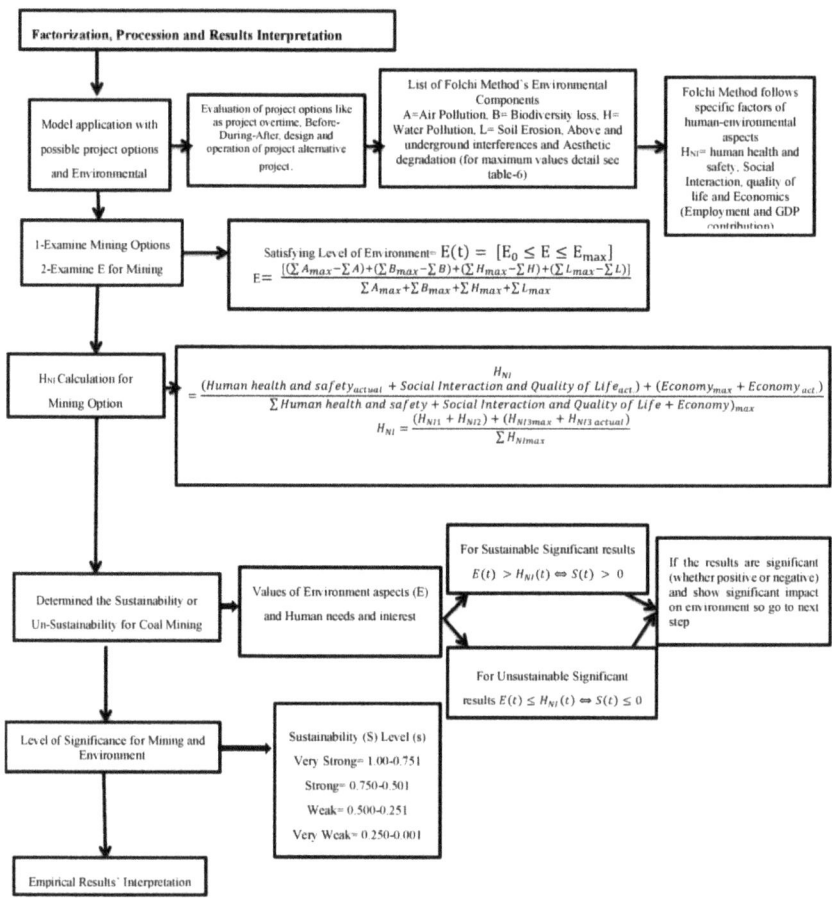

Figure 4. The successive steps of the Folchi Method for measuring environmental sustainability.

Table 2 shows the magnitude of environmental, social, and economic factors. The expected ranges of impacting factors are defined, i.e., intensive impact, medium impact, and low impact. The scores are allotted according to the actual contribution of the factor in the environment. Information will be collected of minors of the Sindh Engro coal mine and residential areas of it.

The correlation matrix of affecting factors and magnitude of components is presented in Table 3. The correlation matrix replicates the endangered level and contribution of environmental factors. The possible contribution of the individual category or factor is divided into three portions, i.e., Maximum (Max), Med (Median), and Minimum (Min). The category and content with no significant relationships or irrelevant are mentioned with zero (Nil) in Table 3. The possible contribution of individual category or factor is divided into three portions, i.e., Maximum (Max), Med (Median), and Minimum (Min). The category and content with no significant relationships or irrelevant are mentioned with zero (Nil) in Table 3.

3.2.2. Phillips Environmental Sustainability Mathematics (PESM) Model

The mathematical model of Phillips is used to examine the sustainable or unsustainable effect of coal mining on the environment by using magnitude recognition values of Folchi.

The model has two parts, i.e., environment and human needs [38,39]. The environmental contribution is measured by following Equation (1):

$$E = \frac{[(\sum A_{max} - (A_1 + A_2)) + (\sum B_{max} - \sum B) + (\sum H_{max} - \sum H) + (\sum L_{max} - (L_1 + L_2 + L_3 + L_4))]}{\sum A_{max} + \sum B_{max} + \sum H_{max} + \sum L_{max}}$$
$$E = \frac{[(\sum A_{max} - \sum A) + (\sum B_{max} - \sum B) + (\sum H_{max} - \sum H) + (\sum L_{max} - \sum L)]}{\sum A_{max} + \sum B_{max} + \sum H_{max} + \sum L_{max}} \quad EQ-1. \quad (1)$$

In the above equation, four main parts of the environment $E(t) = (A + B + H + L)$ are indicated with different abbreviations: "A" for Atmosphere, "B" for Biosphere, "H" for Hydrosphere, and "L for Lithosphere. The fluctuation of spheres is measured concerning time. The maximum safe environmental anthropology system with its maximum limits and sub-spherical operation is examined by $E(t) = [E_0 \leq E \leq F_{max}]$. E denotes the spherical progression, adaptation, renewal, and repairing environment concerning time t. The level of the environment is calculated as $E = \frac{E_{max} - \sum E}{\sum E_{max}}$; the actual contribution of environmental factors is deducted from maximum environmental value and divided by the maximum value of the environment.

The second part of the model is a human need, which presented below in Equation (2):

$$H_{NI} = \frac{(H_{NI1} + H_{NI2}) + (H_{NI3max} - H_{NI3\ actual})}{\sum H_{NImax}}$$
$$H_{NI} = \frac{(\sum H_{NImax} - \sum H_{NI})}{\sum H_{NImax}}. \quad (2)$$

In Equation (2), H_{NI} covers human needs or socio-economic factors, i.e., the collection of (Human Health and Safety + Social Interaction and Quality of Life + Economy), H_{NI1}, H_{NI2}, H_{NI3} indicate three components, respectively. The human needs components $H_{NI}(t) = [H_{NI} \leq H_{NI} \leq H_{NI}\ max]$ indicate the resources and environmental contribution for human life and survival.

After calculating the environmental and socio-economic factors, the next process is to check the relationship between environment, socio-economic conditions, and coal mining in Sindh. This model's last step elucidated the viable working environment and sustainability level in coal mining. To determine sustainability, the calculated human interest and needs value is deducted from the calculated environmental value $S(t) = E(t) - H_{NI}(t)$. If $E(t) > H_{NI}(t) \Leftrightarrow S(t) > 0$, so the sustainable and positive impact on the environment will be declared. If $E(t) \leq H_{NI}(t) \Leftrightarrow S(t) \leq 0$, a negative and unsustainable impact on the environment will be declared.

4. Results

The coal mines are used to select individual environmental components' specific magnitude. As per the Folchi method's assumption, the maximum score of the individual component is 100. The actual weightage of affecting determinants declares the individual contribution to environmental sustainability. Table 3 is generated by multiplying each individual factor's magnitude and correlation matrix (see Figure 5). The first factor, land dispossession and potential resources, received 8 points because the Sindh Engro coal mining area is in the Mithi region's rural areas. The affected population of Sindh Engro coal mining is 12 villages. The exposition and visibility of the pit received 6 points as per its magnitude. The open-pit area can be seen from the road easily. The water pollution and resources of underground and above-ground received 8 and 9 points, respectively.

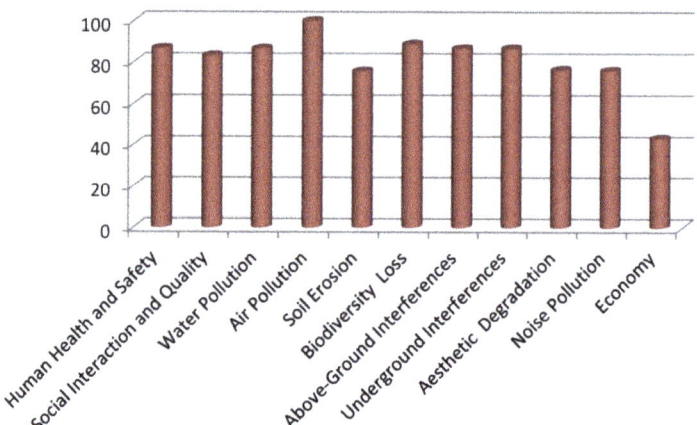

Figure 5. Final weighted impact scores of Sindh Engro coal mining.

The coal mine has affected surface water, facing physicochemical and biological issues. The coal mining exposed rocks containing high sulfur-bearing mineral pyrite and dissolved it with water wash to attain the coal. The sulfur-bearing mineral pyrite reacts with water and air and becomes the reason for acidic water outflow. This compounded form of dilute acid is in nearby rivers, lakes, and streams. The quality of water, underground, and above-ground resources directly endangered human life by its processed, acidic mine drainage and coal sludge refinery process. The coal mining started at Sindh Engro coal mining station in 2015, and the traffic rate increased by 200% due to mining. The increase in vehicle traffic attained 9 points as per its increasing level. Coal mining strongly contributes to the atmospheric release of gases and coal dust, while 10 points are allotted because of the free emission of toxic gases [38–40].

Coal mining diffuses sulfur dioxide (SO_2), nitrogen oxide (NO_2), methane (CH_4), carbon dioxide (CO_2), particulate matter (PM2.5, PM10), and heavy metals in the air. The diffusion of toxic gases and coal dust increases smog, acidic rains, and deadly respiratory, cardiovascular, and cerebrovascular issues. The mine tailing spill attained 9 points because of the design blast and no clearance procedure. The minerals are mined, and due to a lack of resources, the wastages are not disposed of appropriately. The coal fire generally happens during coal mining and coal wastage, so coal fire emits mercury and CO_2. The local environment is facing toxic air, water, and rains due to the mine-tailing spill. The wastage of the mining process disposed of in surface impoundments affected the land's internal and external surface by leaching out the toxic metals [41–43].

The working station's level of noise is between 131 and 141 decibels (DB). That is why the magnitude table allotted 8 points to the noise level. According to the noise chart, the heavy traffic, machinery, and blasting voice are between 131 and 141 decibel levels. The ground vibration received 8 points because of cosmetic damages above the threshold. The working station is in vibrating mode during operation, and the vibration is intense, which can affect minors' body organs. The last two components are related to the economy, i.e., the livelihood of local forces and contribution to GDP. The weightage of local details' livelihood is five because of the medium employment level of residents. The Chinese Engineering Company invested in this mining project, so the engineering team is not local.

The contribution in GDP received 3 points because of low contribution. The Sindh Engro Coal mine is not fully explored yet; it is in the working process. After the multiplication of assigned points to individual component and correlation matrix, the total points of human health and safety, social interaction and quality of life, water pollution, air pollution, soil erosion, biodiversity loss, above-ground interferences, underground interferences, aesthetic degradation, noise pollution, and economy are 86.8, 83.16, 86.58, 100, 75.71, 88.83,

86.66, 86.67, 76.41, 76, and 43.34 respectively. The individual category's summation is used in the Phillips Environmental Sustainability Mathematics (PESM) model to measure the sustainable or unsustainable relationship between coal mining and environmental sustainability.

Table 5 contains the abbreviation of categories and the maximum weightage of individual categories. As per methodological explanation, the modeling covers two main affected areas of coal mining, i.e., environment and human needs and interest [39], [44]. Environmental sustainability is based on four basic parts: atmosphere, biosphere, hydrosphere, and lithosphere. The atmosphere covers air and noise pollution, and the maximum weightage of the atmosphere is 200 (100 for air pollution and 100 for noise pollution). The biosphere consists of biodiversity loss (maximum weightage 100), as the livestock is dependent on trees and the Rohiro trees are banned in operational areas, which threatened the gugrall (Camiphera Mukul), phoge (Clligonum polygonoides), rohiro (Tecoma undulate), Peeloo (Salvadora persica), Kandi (Prosopis cineraria), and Kombhat (Acacia Senegal) species. The hydrosphere is related to water sustainability, which has 100 maximum weightages for this research. The lithosphere is the crust and mental shell of earth and abbreviated with L in the model. The soil erosion, above and underground interferences, and aesthetic degradation cover under the lithospheric part of the environment with a maximum of 400 weights. The second part of the division depends on human needs and interests with 300 weights. The areas covered under human needs and interests are human health and safety, social interaction and quality of life, and economy.

Table 5. Determination of sustainability (S) components.

	Components of Environment (E)	Maximum Possible Scores for E and H_{NI}
A_1	Air Pollution	A max = 2 × 100 = 200
A_2	Noise Pollution	
B_1	Biodiversity Loss	B max = 1 × 100 = 100
H_1	Water Pollution	H max = 1 × 100 = 100
L_1	Soil Erosion	L max = 4 × 100 = 400
L_2	Above-Ground Interferences	
L_3	Underground Interferences	
L_4	Aesthetic Degradation	
Components of H_{NI}		
$H_{NI}1$	Human Health and Safety	H_{NI} max = 3 × 100 = 300
$H_{NI}2$	Social Interaction and Quality of Life	
$H_{NI}3$	Economy	
Evaluation of Sustainability for Sindh Engro Coal Mining		
$S = E - H_{NI}$		

Table 6 shows the final results of environmental sustainability. The first column contains the actual score of environments. In contrast, the second column includes a number of human needs and interests. The third column shows sustainability values with a negative sign. The negative sign of sustainability amount declared that the Sindgh Engro coal mines negatively influenced the environment.

$$E = \frac{[(\sum A_{max} - \sum A) + (\sum B_{max} - \sum B) + (\sum H_{max} - \sum H) + (\sum L_{max} - \sum L)]}{\sum A_{max} + \sum B_{max} + \sum H_{max} + \sum L_{max}}$$

$$E = \frac{[(\sum A_{max} - (A_1 + A_2) + (\sum B_{max} - \sum B) + (\sum H_{max} - \sum H) + (\sum L_{max} - (L_1 + L_2 + L_3 + L_4)]}{\sum A_{max} + \sum B_{max} + \sum H_{max} + \sum L_{max}}$$

$$E = \frac{[(200 - (100 + 76) + (100 - 88.83) + (100 - 86.58) + (400 - (75.71 + 86.66 + 86.67 + 76.41)]}{200 + 100 + 100 + 400}$$

$$E = \frac{[(200 - (176) + (11.17) + (13.42) + (400 - (325.45)]}{800}$$

$$E = \frac{(24 + 11.17 + 13.42 + 74.55)}{800}$$

$$E = \frac{123.14}{800}$$

$$E = 0.153925$$

$$H_{NI} = \frac{(H_{NI1} + H_{NI2}) + (H_{NI3max} - H_{NI3\ actual})}{\sum H_{NImax}} \quad (3)$$

$$H_{NI} = \frac{(86.8 + 83.16) + (100 - 43.34)}{300}$$

$$H_{NI} = \frac{169.96 + 56.66}{300}$$

$$H_{NI} = \frac{226.62}{300}$$

$$H_{NI} = 0.7554$$

$$S = E - H_{NI}$$

$$S = 0.153925 - 0.7554$$

$$S = -0.601475$$

$$E > H_{NI} = \text{Sustainable relation};\ E \leq H_{NI} = \text{un} - \text{sustainable relation}$$

$$E\ (0.153925) < H_{NI}(0.7554) = \text{un} - \text{Sustainable}$$

Table 6. Environmental sustainability.

Mine	E	H_{NI}	S-Value	S-Level
Sindh Engro Coal Mining	0.153925	0.7554	−0.601475	Strong

Environmental sustainability and Sindh Engro Coal mining by employing the Phillips model are displayed in Table 6. The environment (E) value is 0.153925, and the value of human need and interest (H_{NI}) is 0.7554. Therefore, as per the sustainability check, the human needs and interest value is more significant than environment E (t) < H_{NI} (t) ⇔ S (t) < 0. Thus, an unsustainable relation between the environment and coal mining is confirmed, or we can say that Sindh Engro coal mining negatively influences environmental sustainability [12,13,40,45,46].

5. Discussion

The analytical part of this study is based on two sections: a qualitative part (Folchi method) and quantitative part (Phillips Environmental Sustainability Mathematics (PESM) Model). Due to a combination of qualitative and quantitative features, this research confirmed its semi-quantitative quality. The Folchi method covered almost all environment influenced factors as per four significant spheres of environment. Table 1 presents all possible coal mining environments and impacting factors specification. Table 2 contains the numbering according to the intensity of influencing factors for the indexing process. This table also explains the measuring scale of the individual factor for the convenience of readers. Table 3 is most important to understand for interlinking after and before processes.

Table 3 is technically maintained with possible maximum, medium, and minimum points by Folchi. Finally, Table 4 presents the multiplication results.

The aggregate scores of an individual factor in Table 4 declare the actual contribution of these factors in the environment for the Sindh Engro coalfield. The second part shows the result of the Phillips Environmental Sustainability Mathematics (PESM) Model. The expected scores and actual scores are put in the Phillips Environmental Sustainability Mathematics formula. Figure 4 describes the sustainable or unsustainable contribution of coal mining to environmental sustainability. As per our analysis, the environmental contribution is less than human needs and interests. The Sindh Engro coal mines confirmed an unsustainable relationship with the environment. The Sindh Engro coal mines have the largest deposit of coal and explore coal with foreign collaboration. Due to foreign association to digging coal, this coalfield utilized well-developed technology to explore the coal deposits. The unsustainable environmental contribution can ease the way to explore less developed coal mines and eradicate the emissions-friendly resources that will enable the sustainable contribution in the environment and economy.

6. Conclusions

The contribution of coal mining to economic development cannot be denied, but its source of environmental pollution, socio-economic issues, soil erosion, and ecological destruction. In the current study, the Folchi weightage of influencing determinants identifies environmental factors' intensity and contribution. The suitable assessment process is the basis of accurate results and understating the direction of mining impacts. The adversity of environmental determinants is recognized at the planning stage and set the corrective measures. The various coal mining operational and post-operational activities are destroying the environment's sustainability. The correlation matrix is generated by employing the Folchi method, and the final values of the Folchi method are used in Phillips Environmental Sustainability Mathematics (PESM) Model. The research confirms an unsustainable relation between the environment and Sindh Engro coal mining ($E \leq H_{NI}$ = unsustainable relation). Coal mining is increasing water pollution, air pollution, and noise pollution. The underground and above-ground resources face the problem of toxicities. Soil erosion, soil degradation, and soil loss are also going upward due to insalubrious coal mining operations.

The worker's health and safety are entirely ignored in the mining process, and no training sessions are held for new recruiters. As a result, the life of biodiversity is threatened by coal mining in the Thar Desert. The health and safety awareness among workers is poor and becomes the reason for causalities, fatalities, and accidents. The direct discharge of wastages can resolve thar effluents from all blocks into Left Bank Outfall Drain (LBOD) instead of new ponds construction. The studies should analyze the direct and indirect impact of saline water storage, contaminant/solute transport, and evaporation.

The salt concentration tolerance should be checked to utilize dewatered water for bio-saline agriculture and marked salt-tolerant crops and others. Environmental sustainability and biodiversity can be protected by expanding the irrigated agriculture area. Greenhouse cultivation should be promoted because it could reduce the metal-ingestion-associated health risks from edible vegetable biomass and the soil to plant bioaccumulation (BCF). By the higher level of soil organic matters and faster growth rate, the greenhouse cultivation metal uptake rate is less than open fields. Therefore, greenhouse cultivation can provide less contaminated organic food for residents in adverse polluted areas.

The aquatic habitats can be saved by promoting aqua farming at Gordano dam/pond. The extracted groundwater from mines can be transformed into drinking water by Reverse Osmosis (RO). This transformation can reduce the inflow of water into the Gordano dam/pond. The local worker should prioritize job opportunities in the Thar coalfield compared to foreigners. The landholders of Thar who lost their earning due to the construction of coal plants should be considered for mining jobs and projects. The planting of trees, shrubs, and grass should be encouraged along the roadsides and Gordano dam/pond embankments. The planting can reduce wind erosion and silting of the pond, while the local environment will be better. Supercritical and ultra-supercritical coal-fired power plants should minimize gas emissions, i.e., carbon dioxide, sulfur, and mercury, in the firing process. The diminishing rate of toxic gases emission will increase environmental sustainability and decrease fuel costs per megawatt.

This research enlightened the influencing factor of environmental sustainability during the exploration of the coal deposit. Being a developing country, Pakistan is using old methods of coal mining. The environmentalists have expressed their concerns about greenhouse gas reduction and stated that zero greenhouse gas is required within three decades. Practitioners can obscure the problems of irrationalities, less technological development, political instability, and injustices of extraction projects. The specific location of the Sindh Engro coal mines is a developing place, and more specifically, the people of that area are not financially strong. The poor population is focused on getting basic needs and avoiding environmental sustainability policies. The status of residents' lives can be raised by increasing the employment level and sustainable environment awareness. The adoption of green technology during mining exploration can eradicate emissions and positively contribute to environmental sustainability. Developed technology also helps obtain pure refined coal as compared to raw coal. Raw coal emits more carbon dioxide than refined coal. The mining industry management should follow the local and global sustainable environmental policies. As per our findings, the local employment is very low in the Sindh Engro coal mine residents.

Author Contributions: Conceptualization, M.M., S.N.; methodology, M.S., L.I.; investigation, Q.Z.; resources, M.S.; writing—original draft preparation, M.S., L.I.; writing—review and editing, M.S., M.M., Q.Z., S.N.; visualization, M.M.; supervision, M.S., L.I., and S.N.; project administration, M.M. All authors have read and agreed to the published version of the manuscript.

Funding: No external funding is received for this research.

Institutional Review Board Statement: Not applicable.

Informed Consent Statement: Not applicable.

Data Availability Statement: Data are personally collected by structure format of Folchi from the high authorities as well as workers of Sindh engro coal mine.

Acknowledgments: This work is supported by the construct program of applied characteristic discipline "Applied Economics" Hunan Province.

Conflicts of Interest: The authors declare no conflict of interest.

References

1. Ohara, T.; Akimoto, H.; Kurokawa, J.; Horii, N.; Yamaji, K.; Yan, X.; Hayasaka, T. An Asian emission inventory of anthropogenic emission sources for the period 1980–2020. *Atmos. Chem. Phys.* **2007**, *7*, 4419–4444. [CrossRef]
2. Lu, Y.; Wang, Q.; Zhang, X.; Qian, Y.; Qian, X. China's black carbon emission from fossil fuel consumption in 2015, 2020, and 2030. *Atmos. Environ.* **2019**, *212*, 201–207. [CrossRef]
3. Lin, B.; Raza, M.Y. Coal and economic development in Pakistan: A necessity of energy source. *Energy* **2020**, *207*, 118244. [CrossRef]
4. Kousar, S.; Rehman, A.; Zafar, M.; Ali, K.; Nasir, N. China-Pakistan Economic Corridor: A gateway to sustainable economic development. *Int. J. Soc. Econ.* **2018**, *45*, 909–924. [CrossRef]
5. Farooqui, M.A.; Aftab, S.M. China-Pakistan Economic Corridor; Prospects and Challenges for Balochistan, Pakistan. *IOP Conf. Ser. Mater. Sci. Eng.* **2018**, *414*, 012046. [CrossRef]
6. Skoczkowski, T.; Bielecki, S.; Kochański, M.; Korczak, K. Climate-change induced uncertainties, risks and opportunities for the coal-based region of Silesia: Stakeholders' perspectives. *Environ. Innov. Soc. Transit.* **2020**, *35*, 460–481. [CrossRef]
7. Watts, J. We Have 12 Years to Limit Climate Change Catastrophe, Warns UN. *The Guardian*. 2018, p. 8. Available online: https://www.theguardian.com/environment/2018/oct/08/global-warming-must-not-exceed-15c-warns-landmark-un-report (accessed on 16 November 2020).
8. Burke, A.; Fishel, S. A coal elimination treaty 2030: Fast tracking climate change mitigation, global health and security. *Earth Syst. Gov.* **2020**, *3*, 100046. [CrossRef]
9. Sohoo, I.; Ritzkowski, M.; Kuchta, K.; Cinar, S.Ö. Environmental Sustainability Enhancement of Waste Disposal Sites in Developing Countries through Controlling Greenhouse Gas Emissions. *Sustainability* **2020**, *13*, 151. [CrossRef]

10. Rashid, M.I.; Benhelal, E.; Rafiq, S. Reduction of greenhouse gas emissions from gas, oil, and coal power plants in Pakistan by carbon capture and storage (CCS): A Review. *Chem. Eng. Technol.* **2020**, *43*, 2140–2148. [CrossRef]
11. Shen, M.; Huang, W.; Chen, M.; Song, B.; Zeng, G.; Zhang, Y. (Micro) plastic crisis: Un-ignorable contribution to global greenhouse gas emissions and climate change. *J. Clean. Prod.* **2020**, *254*, 120138. [CrossRef]
12. Mohsin, M.; Naseem, S.; Zia-ur-Rehman, M.; Baig, S.A.; Salamat, S. The crypto-trade volume, GDP, energy use, and environmental degradation sustainability: An analysis of the top 20 crypto-trader countries. *Int. J. Financ. Econ.* **2020**. [CrossRef]
13. Sarfraz, M.; Mohsin, M.; Naseem, S.; Kumar, A. Modeling the relationship between carbon emissions and environmental sustainability during COVID-19: A new evidence from asymmetric ARDL cointegration approach. *Environ. Dev. Sustain.* **2021**, 1–19. [CrossRef]
14. Ghose, M.K.; Majee, S.R. Assessment of the impact on the air environment due to opencast coal mining—An Indian case study. *Atmos. Environ.* **2000**, *34*, 2791–2796. [CrossRef]
15. Ahmad, A.; Hakimi, M.H.; Chaudhry, M.N. Geochemical and organic petrographic characteristics of low-rank coals from Thar coalfield in the Sindh Province, Pakistan. *Arab. J. Geosci.* **2014**, *8*, 5023–5038. [CrossRef]
16. Sanjrani, M.A.; Memon, I.H.; Awan, B.A. Environmental Impact of Lakhra Coal Mining, Sindh province, Pakistan. *N. Am. Acad. Res.* **2018**, *1*, 72–75.
17. Cao, C.; Chen, X.-P.; Ma, Z.-B.; Jia, H.-H.; Wang, J.-J. Greenhouse cultivation mitigates metal-ingestion-associated health risks from vegetables in wastewater-irrigated agroecosystems. *Sci. Total. Environ.* **2016**, *560–561*, 204–211. [CrossRef]
18. Hussain, M.; Muhammad, S.; Malik, R.N.; Khan, M.U.; Farooq, U. Status of heavy metal residues in fish species of Pakistan. *Rev. Environ. Contam. Toxicol. Vol.* **2014**, *230*, 111–132. [CrossRef]
19. Obiora, S.C.; Chukwu, A.; Davies, T.C. Heavy metals and health risk assessment of arable soils and food crops around Pb–Zn mining localities in Enyigba, southeastern Nigeria. *J. Afr. Earth Sci.* **2016**, *116*, 182–189. [CrossRef]
20. Zheng, N.; Wang, Q.; Zhang, X.; Zheng, D.; Zhang, Z.; Zhang, S. Population health risk due to dietary intake of heavy metals in the industrial area of Huludao city, China. *Sci. Total. Environ.* **2007**, *387*, 96–104. [CrossRef]
21. Khan, A.A.; Khan, W.A.; Chaudhry, A.A. Mammalian diversity in thar desert habitat of tharparkar district, Sindh, Pakistan. *Pak. J. Zool.* **2015**, *47*, 1205–1211.
22. Masih, A. Thar Coalfield: Sustainable Development and an Open Sesame to the energy security of Pakistan. *J. Phys. Conf. Ser.* **2018**, *989*, 012004. [CrossRef]
23. Ansari, K.A.; Mahar, A.R.; Malik, A.R.; Sirohi, M.H.; Saand, M.A.; Simair, A.A.; Mirbahar, A.A. Impact of grazing on plant biodiversity of desert area of district Khairpur, Sindh, Pakistan. *J. Anim. Plant Sci.* **2017**, *27*, 1931–1940.
24. Zaigham, N.A.; Ahmad, M.; Hisam, N. Thar rift and its significance for hydrocarbons. In *Pakistan Association of Petroleum Geoscientists (PAPG), Islamabad Google Scholar*; Pakistan Association of Petroleum Geoscientists: Islamabad, Pakistan, 2000.
25. Zaigham, N.A.; Nayyar, Z.A. Renewable hot dry rock geothermal energy source and its potential in Pakistan. *Renew. Sustain. Energy Rev.* **2010**, *14*, 1124–1129. [CrossRef]
26. Phuong, T.T.; Shrestha, R.P.; Chuong, H.V. Simulation of soil erosion risk in the upstream area of Bo River watershed. In *Redefining Diversity & Dynamics of Natural Resources Management in Asia*; Elsevier: Amsterdam, The Netherlands, 2017; Volume 3, pp. 87–99.
27. Dondini, M.; Abdalla, M.; Aini, F.K.; Albanito, F.; Beckert, M.R.; Begum, K.; Brand, A.; Cheng, K.; Comeau, L.-P.; Jones, E.O. Projecting soil C under future climate and land-use scenarios (modeling). In *Soil Carbon Storage*; Elsevier: Amsterdam, The Netherlands, 2018; pp. 281–309.
28. Widanarko, B.; Legg, S.; Stevenson, M.; Devereux, J.; Jones, G. Prevalence of low back symptoms and its consequences in relation to occupational group. *Am. J. Ind. Med.* **2013**, *56*, 576–589. [CrossRef]
29. Vearrier, D.; Greenberg, M.I. Occupational health of miners at altitude: Adverse health effects, toxic exposures, pre-placement screening, acclimatization, and worker surveillance. *Clin. Toxicol.* **2011**, *49*, 629–640. [CrossRef]
30. Viljoen, D.A.; Nie, V.; Guest, M. Is there a risk to safety when working in the New South Wales underground coal-mining industry while having binaural noise-induced hearing loss? *Intern. Med. J.* **2006**, *36*, 180–184. [CrossRef]
31. Md-Nor, Z.; Kecojevic, V.; Komljenovic, D.; Groves, W. Risk assessment for loader-and dozer-related fatal incidents in US mining. *Int. J. Inj. Contr. Saf. Promot.* **2008**, *15*, 65–75. [CrossRef]
32. Paul, P.S.; Maiti, J. The role of behavioral factors on safety management in underground mines. *Saf. Sci.* **2007**, *45*, 449–471. [CrossRef]
33. Panhwar, S.; Mahar, R.B.; Abro, A.A.; Ijaz, M.W.; Solangi, G.S.; Muqeet, M. Health and safety assessment in Lakhra coal mines and its mitigation measures. *Iran. J. Heal. Saf. Environ.* **2017**, *4*, 775–780.
34. Brnich, M.J.; Kowalski-Trakofler, K.M.; Brune, J. Underground Coal Mine Disasters 1900–2010: Events, Responses, and a Look to the Future. *Extr. Sci. Century Min. Res.* **2010**, *363*. Available online: https://www.cdc.gov/NIOSH/Mining/UserFiles/works/pdfs/ucmdn.pdf (accessed on 26 December 2020).
35. Grayson, R.L.; Kinilakodi, H.; Kecojevic, V. Pilot sample risk analysis for underground coal mine fires and explosions using MSHA citation data. *Saf. Sci.* **2009**, *47*, 1371–1378. [CrossRef]
36. Naseem, S.; Fu, G.L.; Mohsin, M.; Rehman, M.Z.; Baig, S.A. Semi-Quantitative Environmental Impact Assessment of Khewra Salt Mine of Pakistan: An Application of Mathematical Approach of Environmental Sustainability. *Min. Met. Explor.* **2020**, *37*, 1–12. [CrossRef]

37. Phillips, J. Applying a mathematical model of sustainability to the Rapid Impact Assessment Matrix evaluation of the coal mining tailings dumps in the Jiului Valley, Romania. *Resour. Conserv. Recycl.* **2012**, *63*, 17–25. [CrossRef]
38. Folchi, R. Environmental impact statement for mining with explosives: A quantitative method. *Proc. Annu. Conf. Explos. Blasting Tech.* **2003**, *2*, 285–296.
39. Phillips, J. Using a mathematical model to assess the sustainability of proposed bauxite mining in Andhra Pradesh, India from a quantitative-based environmental impact assessment. *Environ. Earth Sci.* **2012**, *67*, 1587–1603. [CrossRef]
40. Phillips, J. A mathematical model of sustainable development using ideas of coupled environment-human systems (Invited Article). *Pelican Web J. Sustain. Dev.* **2010**, *6*, 127–142.
41. Kholod, N.; Evans, M.; Pilcher, R.C.; Roshchanka, V.; Ruiz, F.; Coté, M.; Collings, R. Global methane emissions from coal mining to continue growing even with declining coal production. *J. Clean. Prod.* **2020**, *256*, 120489. [CrossRef]
42. Oei, P.-Y.; Brauers, H.; Herpich, P. Lessons from Germany's hard coal mining phase-out: Policies and transition from 1950 to 2018. *Clim. Policy* **2019**, *20*, 963–979. [CrossRef]
43. Jakob, M.; Steckel, J.C.; Jotzo, F.; Sovacool, B.K.; Cornelsen, L.; Chandra, R.; Edenhofer, O.; Holden, C.; Löschel, A.; Nace, T.; et al. The future of coal in a carbon-constrained climate. *Nat. Clim. Chang.* **2020**, *10*, 704–707. [CrossRef]
44. Azapagic, A. Developing a framework for sustainable development indicators for the mining and minerals industry. *J. Clean. Prod.* **2004**, *12*, 639–662. [CrossRef]
45. Rawat, N.S. A study of physicochemical characteristics of respirable dust in an Indian coal mine. *Sci. Total Environ.* **1982**, *23*, 47–54. [CrossRef]
46. Tiwary, R.K. Environmental impact of coal mining on water regime and its management. *Water Air Soil Pollut.* **2001**, *132*, 185–199. [CrossRef]

Article

Implications of Entrepreneurial Intentions of Romanian Secondary Education Students, over the Romanian Business Market Development

Amalia Furdui [1], Lucian Lupu-Dima [2] and Eduard Edelhauser [3,*]

1. Management and Industrial Engineering Department, University of Petroșani, 332006 Petroșani, Romania; amaliafurdui@gmail.com
2. Mining Engineering, Surveying and Construction Department, University of Petroșani, 332006 Petroșani, Romania; lucianlupu@upet.ro
3. Head of the Management and Industrial Engineering Department, University of Petroșani, 332006 Petroșani, Romania
* Correspondence: eduardedelhauser@upet.ro; Tel.: +40-722562167

Citation: Furdui, A.; Lupu-Dima, L.; Edelhauser, E. Implications of Entrepreneurial Intentions of Romanian Secondary Education Students, over the Romanian Business Market Development. *Processes* 2021, 9, 665. https://doi.org/10.3390/pr9040665

Academic Editor: Lucian-Ionel Cioca

Received: 22 March 2021
Accepted: 7 April 2021
Published: 9 April 2021

Publisher's Note: MDPI stays neutral with regard to jurisdictional claims in published maps and institutional affiliations.

Copyright: © 2021 by the authors. Licensee MDPI, Basel, Switzerland. This article is an open access article distributed under the terms and conditions of the Creative Commons Attribution (CC BY) license (https://creativecommons.org/licenses/by/4.0/).

Abstract: The study investigates the Romanian entrepreneurial education training program emphasizing the secondary education student entrepreneurial intents included in technical and professional Vocational Education Training (VET) programs, in order to identify its role in increasing student intention in the process of choosing a career as an entrepreneur among graduates of the vocational and technical Romanian education system. The study research methodology was based on the interpretation of two questionnaires consisting of 23 questions, which were applied to a population of 253 and 159 respondents. The survey period was conducted between 2019 and 2020. The respondents were students from the vocational and technical education system in Romania, mostly from the Central Region of Romania, but the results of the study could be extended to the entire Romanian education system. The data were processed using SPSS software, and the results of the study revealed direct, positive, and significant links between psychological and behavioral traits and entrepreneurial intentions of the student surveyed, moderated by the entrepreneurial education acquired through the school curriculum. These results could also be the basis for developing future policies and programs to encourage entrepreneurial behavior, especially for secondary education students from the Romanian education system, specifically on pre-university education.

Keywords: sustainable process; resource efficiency; entrepreneurship; entrepreneurial intentions; education for sustainable development; innovation

1. Introduction

Over the last decade, entrepreneurship has become an important economic and social topic, as well as a well-known research topic in the world. Entrepreneurship is important because it leads to increased economic efficiency, creates new jobs, and brings innovation to the market. The majority of empirical studies have shown that entrepreneurship can be learned, and education can stimulate young people's intention to become an entrepreneur. This has contributed to a continuous increase in the number of entrepreneurship programs; however, the impact that entrepreneurial education has on the intention to become an entrepreneur has remained largely unexplored. The study of youth entrepreneurship contributes to a better understanding of the process of creating employment opportunities for young people. Against the background of the need to stimulate youth employment, entrepreneurship is an important alternative for negotiating the transition from school to work. However, not many aspects of young people's entrepreneurial attitudes, young people's entrepreneurial motivations, and incentives and barriers are known [1,2].

This study aims to investigate the role of entrepreneurial education programs in the entrepreneurial intention of young people, in order to identify the role that entrepreneurship education has on increasing young people's intention in the process of choosing an entrepreneur career among graduates of vocational and technical education in Romania, Centre Region. It is known that reducing youth unemployment is a key objective for policy-makers in developed or developing countries. Analysis of labor market indicators by age grouping shows the need to create jobs for young people. Youth unemployment represents a significant loss of human capital that could contribute to economic growth. Moreover, economies face the need to create employment opportunities for young people, and entrepreneurship is an important way of entering the labor market; and thus, entrepreneurship could contribute to better use of young people's economic potential in order to reduce poverty. Stimulating entrepreneurship among young people brings multiple economic and social benefits, such as reducing unemployment, promoting social inclusion, improving self-confidence among young people, reducing the risk of juvenile delinquency, and stimulating innovation. Thus, companies must support young people's entrepreneurship because it is a valuable source of new jobs and economic dynamism. However, it should be noted that researchers are cautious in treating young people's entrepreneurship as a solution to economic and social problems.

Entrepreneurship is the process in which individuals become aware that business development is a viable option for them by formulating business ideas, learning needed skills for an entrepreneur, and taking the necessary steps to start and develop a business [3,4]. On the other hand, entrepreneurship is seen in close connection with traits such as initiative, innovation, creativity, risk-taking, and the ability to perform in specific economic and cultural environments. Entrepreneurial behavior among young people is also sensitive to a multitude of factors, including entrepreneurship education, institutional and business environment, and access to financing mechanisms [5,6].

2. Materials and Methods

In this study, the authors conducted an analysis of the impact of entrepreneurial education and entrepreneurial intentions as well as the entrepreneurial ecosystem in VET (Vocational Education Training) in Romania for establishing whether the moderate behavioral traits on entrepreneurship can positively and significantly influence young people's entrepreneurial intention. Entrepreneurial education, as a sustainable process, is the support of innovation, which leads to the efficient use of resources. Entrepreneurship involves change and novelty, and one of the reasons it should be encouraged is that "what is different is not always better, but what is better is always different" [7,8]. A positive attitude towards entrepreneurship means a desire to exploit opportunities, with individuals changing the future of their society by changing their own vision of their societal role. The European Union promotes entrepreneurial spirit as a key factor for competitiveness, which represents the importance of implementing and developing a true entrepreneurial culture at the member state level. Entrepreneurship is the most powerful economic force of mankind because entrepreneurs are people who recognize opportunities where others see chaos or confusion and are aggressive catalysts for market change. Entrepreneurship is more than just creating a business—it is taking risks beyond security and having the tenacity to push an idea through reality. Entrepreneurship is the symbol of tenacity and business achievements, and as such, entrepreneurship is an aggressive catalyst for change in the business world [7,8].

2.1. Theoretical Context for Entrepreneurship and Entrepreneurial Eco-Site

Entrepreneurship began to receive attention with the problems brought about by the global economic and financial crisis, and entrepreneurs became heroes capable of providing a boost to fragile economies. Entrepreneurship cannot be accurately defined, and the multidimensionality and homogeneity of the concept make it very difficult to generalize the conclusions of studies on this concept. The main characteristics of entrepreneurship are innovation, proactivity, and risk-taking, and these positive effects are associated with entrepreneurship, directly or indirectly: sustainable development, the overall growth of business performance and economic efficiency, creation of new jobs, improvement of regional and national competitiveness, more efficient management of economic crises, and recognition and exploitation of opportunities [9].

Both researchers and people involved in economic policymaking and development strategies are increasingly emphasizing the role of entrepreneurship in ensuring the economic growth of a country. Entrepreneurship has come to be associated with economic development and the well-being of nations. This statement is supported by a number of researches in the field, such as those conducted by Fernandez-Guadaño et al. They performed an analysis of the population's skills in the managerial field, starting from the traditional business model, demonstrating the usefulness of increasing the individual level of managerial, entrepreneurial competence. The purpose of this growth is to support individual progress because there is a clear relationship between it and the increase in the level of entrepreneurial competence [10]. Research by Buzu and others in this area, taking an approach in conjunction with the circular economy, provides additional evidence. The authors' study proves that the use of circular economy processes is felt at the individual level. Specifically, there are clear benefits in terms of quality of daily life [11].

The particular importance given to entrepreneurs in their studies on economic growth is explained by the fact that they are the ones who introduce new technologies, promote new products, stimulate the discovery of new resources, and mobilize capital. An example with proven relevance is provided by the study conducted by Briciu et al., who analyzed the influence of the use of innovative technologies in tourism entrepreneurship, proving that this approach leads to improved results in this field. Essentially, the avant-garde technology increases the attractiveness of the tourist offer [12]. Meanwhile, going on an even more interesting path, Bican et al., in agreement with the changes that are taking place worldwide regarding digital transformations, analyzed the implications of these transformations in entrepreneurship. The authors discuss the sustainability of digital entrepreneurship, presenting an extremely favorable perspective on this type of entrepreneurship [13].

Public information amply proves that entrepreneurs are the ones who provide jobs for most of a country's population, which consolidates the middle class. Research on family entrepreneurship presents very clear evidence of the importance of this social segment. Schepers et al. conducted an analysis at the level of family firms that established a clear link between the long-term orientation of the family and entrepreneurship. The authors' study analyzed how the disappearance of family members clearly affected the firm [14]. As a link, in terms of entrepreneurs' attitudes towards the environment and the legislation governing their business, Keiko Yamaguchi et al. looked at family businesses in agriculture, proving the importance of training entrepreneurs [15].

2.1.1. Analysis of the Entrepreneurial Ecosystem

Various definitions for the term entrepreneurial ecosystem can be found in the literature, all of which are that it encompasses the social and economic environment affecting entrepreneurial activity. The concept of an entrepreneurial ecosystem highlights the collective and systemic nature of entrepreneurship. Therefore, in order to stimulate entrepreneurial activity in a particular country, it is necessary to facilitate the creation of an ecosystem conducive to entrepreneurship. The approach of entrepreneurship based on personal attributes is explained by the fact that not everyone is willing for entrepreneurial activity. Carriles-Alberdi et al. analyzed the influence of the entrepreneurial ecosystem and highlighted important aspects, namely that people with higher education have a greater inclination towards entrepreneurship. The authors also found that in countries based on innovation, women are more easily incorporated into the entrepreneurial phenomenon [16]. Also, a large part of potential entrepreneurs never starts entrepreneurial activity, or shortly after starting their business, they abandon it. In order to identify the stage of new entrepreneurs, entrepreneurs are divided into four categories in the Global Entrepreneurship Monitor (GEM) study [17]. In order to bring real input to a country's economic development, it is necessary for newly created businesses to become persistent, to exceed 3.5 years of activity, during which the newly created businesses are unstable and threatened by the risk of discontinuity. At the same time, both the willingness of a country's population to start businesses, as well as the ability to go through the entrepreneurial stages as from potential entrepreneur to entrepreneur with 3.5 years of experience, depend on multiple environmental factors, such as access to financial resources, government policies, tax facilities, government and non-governmental business support programs, entrepreneurship education and training, access of small businesses to Research and Development (R&D) activities, business infrastructure as well as cultural and social norms. In other words, the intensity and persistence of entrepreneurial activity are subordinated to the quality of the entrepreneurial ecosystem characteristics of that country.

Creating an entrepreneurial-friendly ecosystem is an extremely difficult task because the ecosystem is made up of multiple elements. Entrepreneurial ecosystems have become the object of research of governments, development agencies, and academia. Kim et al. have highlighted in their research that it is very useful for educational institutions to create a social entrepreneurial ecosystem. The advantages of such an ecosystem are brought to other entities that need more active interactions [18]. Organizations such as the US Competitiveness Council (COC), the GSM Association, the Organization for Economic Cooperation and Development (OECD), the World Bank, and the World Economic Forum have developed complex tools for assessing and tracking ecosystem development. In addition, there are models for assessing ecosystems developed by investors. The relevant models for evaluating entrepreneurial ecosystems are Babson Entrepreneurship Ecosystem Project—Babson College, Asset Mapping Roadmap—Council on Competitiveness, USA, Global Entrepreneurship and Development Index—George Mason University, Innovation Rainforest Blueprint—Hwang V.H., Six + Six—Koltai Company, Information and Communication Technology Entrepreneurship, Entrepreneurship Measurement Framework—OECD, Doing Business—World Bank, and Entrepreneurship Ecosystem—World Economic Forum [19,20]. Each of these entrepreneurial ecosystem evaluation models is based on multifactorial analysis, e.g., the model developed by Babson University involves investigating the simultaneous state of 13 distinct domains, where at the center of the ecosystem is the entrepreneur, who carries out entrepreneurial activity, but his activity is accelerated or, on the contrary, braked by the ecosystem. Each of the 13 spheres that form the ecosystem generates favorable factors or impediments to entrepreneurship. The entrepreneurial ecosystem in Romania is developing, although the country has a lot of potential, and the dynamics of the changes needed to promote an innovative economy and a digital society are quite small. Expert assessment of the current state of the entrepreneurial ecosystem indicates that the biggest barriers to its development are lack of confidence, predictability, and transparency, as well as access to funding and the limited capacity of educational institutions to play an active role as a stakeholder in the entrepreneurial ecosystem. The entrepreneurial ecosystem in the Babson University approach is relevant for this [21].

In addition, inadequate communication between ecosystem stakeholders and poor coordination of government decision-makers slows down the ecosystem's evolving process. Undoubtedly, there are islands of excellence, so access to a supply of talent with an adequate education is one of Romania's greatest strengths. Bratucu et al. managed to establish a link between regional development and sustainable development of universities through an analysis conducted at Transilvania University of Brașov. This aspect is seen from the purely economic perspective of the aforementioned relationship. The inclusion of entrepreneurial training for students, the creation of business incubators, and student entrepreneurial societies represent, in the opinion of the authors, the chance to develop the university-student ensemble. The result, which is important for the future, is to provide young people with the necessary support to work in the entrepreneurial field [22]. However, brain migration, insufficient funding opportunities, and reduced entrepreneurial culture at the secondary level of education require radical policy measures to make the current state of affairs change more quickly. [23] The COVID-19 pandemic highlighted another important aspect. Barbulescu et al. analyzed the need for business development and restructuring on a sustainable basis in the context of the COVID-19 pandemic. The authors' research allowed them to conclude that it seems necessary to involve entrepreneurs in this restructuring process. It seems that collaboration between companies, the public sector, academia, and society, in a quadruple spiral, can ensure targeted sustainable development [24].

2.1.2. Motivations for Entrepreneurship and Barriers for Young Entrepreneurs

To maintain the research problem on entrepreneurship, a narrow definition of entrepreneurship can be used, which focuses on the objective of generating income, i.e., those people who are entrepreneurs, want to become entrepreneurs, or want to start a new business, either in the formal or informal sector, to generate income.

For this purpose, this study is based on youth entrepreneurship; therefore, the authors did not define a specific age range to focus on for young people, and they keep it as large as possible. This is because the definition of young people's ages varies according to context. For example, official countries and statistics define young people in different ways, so setting an age range causes problems in terms of data comparability. Similarly, different organizations define young people differently, so reducing young people to specific age can render this project inappropriate for some organizations unnecessarily. For example, the YBI defines young people up to 35 years of age, while the UN defines young people between the ages of 15 and 24. Moreover, there are more entrepreneurs in the 25 to 34 age group than any other age range (GEM 2011), which suggests that a wider age range, e.g., if the age of 35 is used as the highest age, will include a fairly large portion of all entrepreneurs and will therefore allow conclusions to be drawn from the general literature of entrepreneurship and from the data for young entrepreneurs.

It is important to recognize that not all entrepreneurs are the same. Therefore, the authors believe that young entrepreneurs will present a variety of individual factors or profiles, which may create different needs for support interventions for young people's entrepreneurship between these groups. Several studies explore different ways of classifying young entrepreneurs. For example, Chigunta (2002) proposes a transitional classification in three age groups: the formative stage of pre-entrepreneurship (15–19 years), the growth stage of budding entrepreneurs (20–25 years), and the primary stage of the emerging entrepreneur (26–29 years). Lewis and Massey [25] provide a more diagnostic framework for young entrepreneurs. As described by Schoof [6], the paper distinguishes four types of young entrepreneurs along a continuum of the level of training, e.g., the level of skills and exposure to the enterprise, the young entrepreneurs (potential), and the level of intention to engage in entrepreneurial activity. These types of models can provide some information about the specific needs of different groups of young entrepreneurs and about how to prioritize appropriate interventions to maximize the impact according to the needs of that group profile; however, the authors need to focus this study on terms of disaggregation.

To get a more specific view of potential young entrepreneurs, Lewis and Massey [25] provided a diagnosis framework for young entrepreneurs. It depends on young people's level of training to engage in business and the level of intention to be an entrepreneur. The authors distinguish four different groups of potential young entrepreneurs. This type of framework can help researchers and those who are policy-makers better understand the special needs of young people as a group and identify appropriate methods and promotion programs in order to improve the entrepreneurial ecosystem in a selected country or region [25]. There are five factors that influence youth entrepreneurial engagement to encourage youth entrepreneurship. This includes the following: social and cultural attitude towards youth entrepreneurship, entrepreneurial education, access to funding or starting funding, administrative and regulatory framework, and business support and assistance [25].

The social and cultural attitude towards youth entrepreneurship refers to cultural and social environments that influence an individual's approach to life and similarly influence entrepreneurship and enterprise culture. Gibb defined an enterprise culture as a set of attitudes, values, and beliefs that operate within a community or environment that lead to entrepreneurial behavior as well as the aspiration of self-employment [26].

Researchers have long realized that cultural attitudes influence the entrepreneurial activities of a population, country, region or ethnic group, and they interact as well. Thus, cultural differences between nations are increasingly understood as important to a nation's level of economic and entrepreneurial development. In a cultural environment in which entrepreneurship is respected and harnessed, business failure is treated as a useful learning experience rather than as a source of stigmatization [27].

Cultural perceptions of entrepreneurship depend on additional factors. First, young people need to have perspective and exposure to social problems in need of successful models. Entrepreneurial education and the competitiveness of a country does not start from the factory or in an engineering laboratory—it begins in the classroom [28]. This is important to help young people develop entrepreneurial skills, attributes, and behaviors and for enterprise development and, most importantly, to understand and achieve entrepreneurship as a career option. Thus, entrepreneurial education is not only a means of encouraging young people to become independent, but it is an opportunity to develop the attitudes, defined as personal responsibility, and skills, defined as flexibility and creativity, needed to cope with the employment barriers in today's societies. Entrepreneurial education becomes a support for environmental aspects. Oncioiu et al. analyzed the possibility of modifying business processes in the green field and regarded the effectiveness of these measures [29]. Another argument is provided by Cárcel-Carrasco et al., who analyzed the influence of industrial maintenance on the strategic planning of companies, concluding that it exists and resides in better economic results [30].

The lack of adequate financing to start a business is one of the biggest barriers and impediments for young people who want to start their own business. The lack of funding was also considered to be a more severe barrier than the administrative framework and administrative regulations. Administrative framework and regulations on youth and business entrepreneurship is generally a fairly new area of research. However, these tasks are among the most important barriers to starting young businesses. Government regulations and bureaucratic formalities are also seen as a reason for the large informal sectors in many developing countries because formalization costs are higher than productivity gains. Entrepreneurs also face numerous administrative burdens, including companies registering, tax administration, obtaining investment approvals and commercial licenses, compliance with copyright and patent regulations, competition law, access to workspace and long-term rentals, building and construction permits, customs clearances, utility connections, etc. Business support and business consulting services are very useful for a young entrepreneur because more business support at the beginning and early stages encourage lasting success. These youth support services are mentors, business clubs, and business incubators. These support services can hold the key for transforming youth start-ups with one person into successful small and medium-sized enterprises [31]. This statement is also supported by research conducted by Ibáñez García et al., who analyzed students' perceptions of training through mentoring programs. The authors found that the benefits are very high, with students highlighting the mutual benefit of mentoring [32].

In conclusion, there are various reasons why young people decide to start a business, in relation to their life circumstances, attitudes, preferences, personal goals, interests, and individual strengths. Recognition of these reasons is crucial for understanding and stimulating young people's entrepreneurship.

2.1.3. Theories of Entrepreneurial Intentions

The first formal theory about entrepreneurship was developed by Richard Cantellon in 1725 [33]. He defined entrepreneurship as those self-employed people who take the risk of buying at certain prices and selling at uncertain prices. Later, in 1803, the definition of entrepreneurship was extended to include factors of production, where an entrepreneur transferred resources from locations where they exist in places where there is a shortage of those resources [34].

None of these definitions explicitly captures the trait Josef Schumpeter considers "sine qua non" (without which you cannot) in becoming an entrepreneur of success, i.e., innovation. In his vision, an entrepreneur is a person who develops new technologies and products that change the environment in which they operate, create new opportunities, and change the way of thinking and operating in a particular field of social activity. It is worth noting that Schumpeter saw the entrepreneur not as an inventor of new products but as an explorer of new opportunities that lead to innovation. Together, these three definitions contribute to a common idea of entrepreneurship: risk-taking, coordination, and innovation [35].

The entrepreneurial intentions of individuals are the most important variables that predict their entrepreneurial behavior. Referring to the entrepreneurial intentions of individuals, the current literature adopts several concepts such as career orientation [36], budding entrepreneurs [37], etc. One of the main theoretical works that guided research on entrepreneurial intent comes from the Theory of Planned Behavior [38], according to which the intentions that precede any type of planned behavior are determined by three factors: attitude towards behavior, subjective norm, and control of perceived behavior. Indeed, most models of entrepreneurial intent rely on Ajzen's model, arguing that both psychological and behavioral variables are important in fixing entrepreneurial intent. Political and economic factors, as well as the social context, such as social support, subjective norms, and the perception of opportunities and resources, can also contribute to the formation of their intention to be their own employee [38]. This aspect of the social context was an argument for García-Jurado et al. The authors analyzed the entrepreneurial field, considered to be an emerging field, under the influence of the millennial generation, much more connected to technological progress. The authors found that this influence takes on the appearance of a phenomenon over the millennia and a growth factor of entrepreneurial intent [39].

Shapero and Sokol's theory of entrepreneurial event argues that bad business is the result of interaction between contextual factors that act through their influence on individual perceptions. There are two types of perceptions: perception of desire that refers to the degree the individual feels attracted to the given behavior (to become an entrepreneur) and perceiving the possibility of realization, which refers to the extent to which people consider themselves personally capable of exhibiting a certain behavior [40]. Shapero and Sokol believed attitude is closely linked to entrepreneurial intent, particularly the perception of feasibility and opportunity, and that attitudes derive from the previous exposure to entrepreneurial activities, including both the scale and positivity of previous activities [40].

The first courses on entrepreneurship began in the United States in the 1940s. Then, entrepreneurship education increased considerably around the world. One of the four strategic objectives of the Strategic Framework for European Cooperation in Education and Training, approved by the Council of the European Union in 2009, is to increase creativity and innovation, including entrepreneurship, at all levels of education and training. In this respect, the basic knowledge acquired through entrepreneurial education involves the following: determining opportunities, formulating opportunity through the creation of new ideas and the mobilization of the necessary resources, building and managing a new business, and developing creativity and critical thinking skills [41].

Drucker considered that entrepreneurship is a discipline [42] and for the bigtime, entrepreneurship plays a key role in improving the skills of potential entrepreneurs and their orientation towards the ability to create, develop, and maintain new businesses. Research in this area has shown that through entrepreneurship programs, young people develop entrepreneurial attitudes and skills. Additionally, different educations and skills can explain why some people are engaged in entrepreneurial activities and enjoy more success than others, and students enrolled in these programs can be future entrepreneurs. It is important that educational support is tailored based on their needs and desires to learn and achieve. About this, Lu et al. analyzed the entrepreneurial support university students received and whether it influenced their entrepreneurial intention. It was found that the perception of students was largely positive, which led the authors to conclude that it is necessary to create a stronger entrepreneurial climate at the university level. It also demonstrates the usefulness of organizing entrepreneurship competitions, in addition to entrepreneurship courses [43].

Education is vital to understand entrepreneurship, to develop entrepreneurial skills, and to contribute to their entrepreneurial identities and cultures on an individual, collective, and social level. The need for entrepreneurship education is a certainty. Butum et al. investigated students' perceptions of the skills they acquire and their perspective of their professional future and concluded that they have very well-defined opinions about their need for knowledge [44]. Therefore, an individual who receives an entrepreneurial education that provides him with the necessary administrative skills will be engaged in entrepreneurial activity in the future. This educational component has additional value, correlated with the professional path of the people. This aspect is revealed by Cioca et al., who studied the level of correlation of the profile of young people enrolled in higher education programs. This aspect is very important because many young people change their field of study along the way, which prejudices the educational system, the state that supports the educational component. There is an explicit need for support in career guidance for young people, as almost a third of students choose careers that do not coincide with their interests [45]. Over time, studies have shown that there is great efficiency in education, leading to an increase in the number of entrepreneurship programs in the education system through various forms. Frolova et al. have also researched ways to increase the effectiveness of entrepreneurial education. The authors' analysis led to the conclusion that a new approach to motivating students is needed. The research is valuable because it generates motivation models that can be applied later in the field of human resources [46]. It has also been observed that entrepreneurial skills and intentions can be easily learned with the help of provided educational support that is based on practice, as Bratucu et al. have shown [22]. In addition, since knowledge consolidation should not be ignored, the study by Yuan et al. proves the usefulness of entrepreneurial education. The authors conducted an empirical analysis comparing entrepreneurs and employees with higher education. The authors found that entrepreneurs have a much greater concern about adding skills to their existing baggage. This conclusion comes with the condition that it is proven they have an additional level of stress as a result of the activity they carry out [47].

2.2. Entrepreneurship in Romania

In recent decades, after the 2008–2009 economic crises, the importance of entrepreneurship for economic growth and development at a global level has been a topic that has captured the interest of researchers and practitioners, due to the multidimensional aspects of the phenomenon, with microeconomic and macroeconomic implications [48]. The European Union's entrepreneurship policy constantly highlights the need to create the most appropriate micro- and macro-environments to support the development of small- and medium-sized companies and entrepreneurship. The opinion on the entrepreneurial ecosystem for the former communist countries is that they are behind other European countries, and the differences between the East and West are obvious when the performance of start-ups is analyzed. However, Romania has a strong entrepreneurial ecosystem despite the low level of economic development. Entrepreneurship is valued in society, and 48% of the working population prefers to develop their own business, according to Global Entrepreneurship Monitor data, compared to the European average of 37%. The preference for entrepreneurship could be explained by the low level of wages in Romania, in which context people have higher potential incomes by going on their own. An important aspect is also offered by the social specifics of Romanians, as can be found from the conclusions of the study conducted by Hatos et al. on the individual approach to entrepreneurship in Romania. The authors conclude that individualistic approaches are not specific to people who can become entrepreneurs [49]. However, half of the Romanian respondents say they do not have the financial resources to develop their own business, more than double that of the European Union, where 21% of the respondents say that lack of money is the reason why they do not develop their own businesses.

The Global Entrepreneurship Monitor (GEM) measures entrepreneurial activity through seven indicators: the rate of entrepreneurship in training, the rate of start-ups, the total entrepreneurial activity (TEA), the stable business rate, the rate of business discontinuation, the rate of entrepreneurship motivated by need, and the rate of entrepreneurship motivated by opportunity. The main indicator used by the GEM to measure entrepreneurial activity in a country is the Total Entrepreneurial Activity (TEA), which is an indicator representing the working population between the ages of 18 and 64 years who are in the process of starting a business (entrepreneurship rate in training–nascent entrepreneurship rates) or already running their own start-ups (rate of newly established companies—new business ownership rates). The TEA indicator for Romania is 9%, one percentage point higher than the European average, which ranks it sixth in Europe in terms of total entrepreneurial activity after Austria, Estonia, Latvia, the Netherlands, and Slovakia. The TEA structure is similar in Romania and the EU, with the rate of entrepreneurship in training representing about 60% of the total entrepreneurial activity [50].

The stable business rate measures the percentage of the common path at cross the start-up threshold and reflects the sustainability of entrepreneurship in a country, and the businesses that reach this level are the ones that innovate and create new jobs. In general, the TEA is high in developing countries, but the newly established business can do not last over time. This is also the case of Romania, which is placed in the penultimate place in the EU, in terms of company sustainability having a stable business rate of 4% for over 50% of entrepreneurial initiatives. In Romania, 25% of entrepreneurial activity is driven by need, and 38% of them takes place to spread the opportunities encountered on the market. The situation is consistent with Romania's status as a developing country; thus, Romania has more entrepreneurs motivated by need compared to the European average, and a lower percentage of people motivated by market opportunities (38% compared to the European average of 47%). In Romania, there is an upward trend in the rate of entrepreneurs in the process of being formed and also in the importance of the entrepreneurial programs and educational support offered by the Romanian education system, as was shown in the rate of trainee entrepreneurs in Romania in the GEM [17].

Compared to other countries of the European Union, this rate is a small one, but it is nevertheless appreciated that the emphasis is on promoting entrepreneurship and supporting young people's educational support in order to be able to move towards a future career as entrepreneurs. Entrepreneurs' aspirations and their prospects for the future of their own businesses are important because they show the degree of reding in the ecosystem in which they operate. International education shows the percentage of TEA for companies that have at least 25% of their clients outside the country's borders. The new product rate reflects how innovative a start-up is and predicts the percentage of a business in TEA that claims that the product or service it offers is none for at least some customers. Finally, the expectations of further growth show a company's future development prospects and are calculated as the percentage of TEA that estimates they will employ at least 5 people over the next 5 years.

Significant improvements were also recorded in the informal investor rate, measured as a percentage of the active population that provided financing for a start-up set up by a third person in the last three years. In the absence of a solid ecosystem to finance new businesses, this rate is important because it shows the willingness of the population to invest in start-ups [51].

Romania is among the first European countries where entrepreneurial intentions are high in the whole society. The positive attitude towards entrepreneurship influences the probability of becoming an entrepreneur and the level of support that new companies will receive, whether it is access to finance, partners, or mentors. So, 71% of Romanians consider entrepreneurship an excellent career alternative compared to only 58% of Europeans. Additionally, 74% of the Romanian population believes that entrepreneurs have a privileged status in society. In Romania, there was a decrease in entrepreneurial activity, and also, after 2015, Romania is no longer found in the GEM reports. This process of declining entrepreneurial activity is most likely caused by combined reasons, which have been unfavorable to the entrepreneurial intent. Statistics at the European level indicate that Romania allows little importance to entrepreneurial education, so that less than 10% of those who have started and developed a business have a theoretical basis for this, compared to the European average of 30%. In this regard, Romania being a member of the European Union. Antohi et al., starting from the need to understand the sustainability of the absorption of European funds, analyzed, among others, the error phenomena that affect the absorption rate. Starting from the audit role, the authors come to establish the relationship between the professional level of entrepreneurs and the sustainability of the absorption of European funds [52,53]. Approaching another perspective, Paunescu et al. analyzed the Romanian entrepreneurial environment to understand the component of entrepreneurial intention. The authors demonstrated that, in Romania, the incentives for entrepreneurship could be offered by the technological availability and stability of the rules [54]. This demonstration, touching on the delicate subject of rule stability, provides another reason for the decline in entrepreneurial activity. It cannot be ignored that the result of research conducted by Costache et al. analyzed the factors that affect Small and Medium Enterprises (SMEs) and correlated with them, the decision factors involved, highlighting the positive and negative aspects. Adaptation to the context is what allows SMEs to be successful in their business. In this way, another reason is revealed, this being the ability to adapt [55].

From Global Entrepreneurship Monitor [17], the authors identify the entrepreneurial behavior and attitudes in Romanian entrepreneurial framework conditions. The GEM study examines the characteristics, motivations, and ambitions of people starting businesses, as well as social attitudes towards entrepreneurship. First, the employment rate of the population aged 20 to 64 in Romania (63.9% in 2013) is lower than the EU average (68.5% in 2012), with a national target of 70% by 2020. Second, problems arise, however, when the sustainability of the Romanian entrepreneurial environment is analyzed in terms of total entrepreneurial activity, with 9% of the active population being in the stage of start-up or pre-start-up, over 50% of the newly established initiatives do not survive the critical period. In this context, the focus should be on the development of entrepreneurship education in the education system, either pre-university, university, and all parts of life.

2.3. Attitudes and Perceptions from the Perspective of Technological Entrepreneurship

The idea of introducing entrepreneurship into education and stimulating entrepreneurship could result in a multitude of effects, such as economic growth, the workplace creating and increasing the social resilience of and from individual growth, school involvement, and improving equality. This aspect is also supported by the results of the research by Manning et al. They researched the importance of including entrepreneurship training programs for agricultural specializations. The authors point out the need to develop a new conceptual spiral pedagogy to improve teaching in agricultural schools. It reaches the training in the eco-entrepreneurial field, and it is very useful as a border element for specialist managers in agriculture [56]. In this regard, the need for regional sustainability has led Hagebakken et al. to study the possibility of using entrepreneurial education to build this sustainability. Although the long-term impact of entrepreneurship education is difficult to determine, the authors prove the usefulness of this type of education through analysis among Nordic start-ups in Norway [57]. The implementation of this idea, however, has posed significant challenges alongside the declared positive effects. A lack of time and resources, fear of teachers changing, prevention of educational structures, evaluation difficulties, and a lack of definitional clarity are some of the challenges practitioners have faced when trying to implement entrepreneurship and technology in education.

The European Commission's Entrepreneurship 2020 Action Plan, published in 2013, mentions the importance of focusing on developing cross-cutting and entrepreneurial skills in order to improve the employability of young people. Technological entrepreneurship does not refer to a single individual or the inventions they introduce. It is about managing common exploration and exploitation, in which each individual has roles and responsibilities in terms of progress, and also in collaboration and cooperation with achieving common objectives [58]. Technological entrepreneurship refers to the investment and execution of firms' projects, not just to the recognition of technology or market opportunities. Entrepreneurship-related skills and perceptions are analyzed by the Global Entrepreneurship Monitor using seven indicators: the perception of business opportunities, the perception of one's own skills, the fear of failure in entrepreneurship as a career alternative, the status given to successful entrepreneurs, and the attention given to entrepreneurship. The authors analyze the situation of these indicators for the 27 European countries, together with the arithmetic average for each of them. From the entrepreneurial attitudes and perceptions in EU countries, Romania registers values higher than the European average for most indicators. This is also supported by the fact that studies have shown there is a direct relationship between the perception of one's own skills and competencies on the one hand and the level of economic development of a country on the other [50].

Romanians are prone to risk, and that is why the fear of failure is higher than the European average (41% compared to 39% at the European level). Romania is the first country in Europe in terms of entrepreneurial intentions because 27% of Romanians claim they want to start their own business. This aspect is also relevant in the context in which Romania is one of the post-transition economies, which was targeted by the study of Voda and others. They conducted research in post-transition economies and demonstrated a relationship between individuals' income levels and entrepreneurial intent. People with higher incomes are the most inclined to start an entrepreneurial activity. The novelty element is conferred by the fact that the previous barrier was overcome, according to which the entrepreneurial intention was discouraged by the existence of some relational connections, or other support elements of the individuals, to encourage this intention [59]. The important thing is to study attitude in any context because attitude could greatly affect a person's behavior, and a model of attitude in relation to behavior is proposed by Fishbein and Arjen [60].

This theory holds that attitudes towards behavior, including that of entrepreneurship, predict intentions that in turn predict real behavior; however, it is debatable to agree that a potential entrepreneur should first develop an attitude towards a new enterprise, and this attitude will lead to the intention to initiate a start-up. If the intention of such an initiative is strong, the entrepreneur takes steps in turning it into action. As it can be seen, the value or expectations held by an individual related to the individual's beliefs about an object or event constitutes the basis for the development of the attitude, either positive or negative, towards the object or event. This is transformed into intentions and, consequently, into behavior towards the event or object, including that of the creation of enterprises [61].

The measures necessary to improve entrepreneurship can be taken at both the public and private levels. As a result of the study, the authors believe that there should be a clear division of responsibilities at the government level, so the state should focus on improving the quality of entrepreneurial education programs and, in this respect, there should be a close link between school and business, and the curriculum should be adapted to market conditions [62–64].

The transition from the traditional innovation system to the new philosophy—open, expanded, and collaborative innovation—has multiple causes. This transformation is the result of changes that occurred in the business environment and in society, while the airing of competition and accelerating technical, scientific, and social progress leads to technological entrepreneurship. Among the factors that have contributed to this, new concepts in innovation can be remembered: the increase in R&D spending, with the effort being difficult to sustain even by large companies, increasing complexity of the problems facing humanity, requiring collaboration and convergence of forces to reduce risks, and increasing R&D's completion speed of development and innovation projects [65].

2.4. Entrepreneurship in Vocational and Technical Education in Romania

Since 2009 the Council of the EU, where the Council of the EU and representatives of the Governments of the Member States on the development of education in a fully functional knowledge triangle, have proposed an attitude for promoting a creative, innovative and entrepreneurial perspective among students, teachers, and researchers, which would support the progressive development of a broader entrepreneurial culture through vocational education and training, as well as a more dynamic European labor market and a more skilled workforce. In an educational system that is meant to be modern and in close connection with the needs of a society, the pre-university education level should keep up with all the novelty elements required by this evolution. The teaching method, approved by the Romanian Ministry of Education (MEC) in partnership with the National Center for the Development of Vocational and Technical Education (CNDIDT), on the exercise firm implemented at the level of vocational and technical education in Romania aims to facilitate the transition from school to the active life of students and to integrate students more easily to the future status of employee or employer of a company simulating the processes of a real company [66]. Students and coordinating teachers will work together in virtual firms, firms created by students and monitored by those authorized to run the program, the CNDIDT structure. For this, the correlation between innovation and entrepreneurship becomes mandatory. In this regard, Anser et al. have proven that technological innovation is a relevant aspect in supporting the entrepreneurial component. Innovation as a component of technological education is the element that supports the achievement of efficiency in the correct development of processes in an organization [67].

In this context, a broad, modern approach to the concept of entrepreneurial education and entrepreneurial skills should be connected to today's social needs, and this must analyze entrepreneurial education in terms of status, skills, content, and values in order to advance the proposals on improving entrepreneurial skills in Romanian vocational and technical education (VET) [68]. In order to establish an accurate picture of entrepreneurial education in Romania, in terms of concept and status, the authors have analyzed relevant curricular documents such as elementary, secondary, and high school education plans and curricula from secondary and high school subjects at the national level. After analyzing all the elementary, secondary, and high school study plans, the authors found that entrepreneurship education does not appear as a separate topic in the curriculum, either at the primary or secondary level, even though it is the subject that could best contribute to the realization of one of the eight key European competencies with the aim of initiative and entrepreneurship [69]. Students learn entrepreneurship education only in grade 10, one hour a week, and in grades 11 and 12 in technological high schools only, while in the service profile, they still study entrepreneurship through the interactive method of the exercise firm. With this method, the practice firm was meant to train the beneficiaries practically in the field of information processes, broadening their field of knowledge in a field of vocational training, which the mandatory curriculum does not provide for this activity. Through the vocational evaluation and advice program, as well as through the specific activities of the exercise firm, students identify optimal areas of professional development and acquire basic skills in the field of commercial, financial-accounting, human and legal resources activities, as well as teamwork. The use of these tools increases the success rate in employment and the creation of new jobs. The results obtained by using these tools are affected by the media's influence. Regarding this component, Yasir et al. showed through their study that the media's consequences on attitudes towards sustainable entrepreneurship. The authors clearly demonstrated the importance of supporting talent in this field and increasing entrepreneurial skills [70].

2.5. Entrepreneurship Intentions in the Romanian Business Market

The 2020 COVID-19 crisis had a terrible impact on the number of newly created companies, producing a decreased of 60% between March and May compared to the previous situation. Therefore, 2020 has a very good chance of being the year with the fewest companies established in the last 15 years for Romania, but probably for Europe as a whole. From an average of 12,000 to 14,000 new companies per month, April 2020 brought only 2500 newly created companies.

That is why the authors have used the 2019 data. The number of Romanian Small and Medium Enterprises (SMEs) at the end of 2019 was 723,604, where the number of SMEs from the Center region represents only 11% of the total number [71].

However, from the whole number of Romanian SMEs, an important percent is represented for the Romanian economy by the startup companies created with non-refundable financing. These are companies created during the last ten years, but especially in 2017–2019, financed through two different programs called Start Up Nation and Romania Start Up Plus, first financed from the Romanian budget and the second from the EU budget. Both programs offer around 40,000 non-refundable euros, which represent 200,000 lei.

Start-Up Nation Romania (SUNR) was a great success, given the fact that in the competition, 18,000 companies applied in 2017, respectively, 33,000 companies in 2019, and more than 8500 companies were beneficiaries in 2017, and in 2019 approx. 10,000 companies received funding. Therefore, it is considered that in the world of non-reimbursable funds, Start Up Nation has aroused the greatest interest in the last 10 years [72].

The second entrepreneurship scheme financed by Human Capital Operational Program (POCU), called Romania Start Up Plus (RSUP), was also a success, with over 8700 new businesses were set up through these financing lines, within the 205 projects implemented by the grant administrators selected by the Ministry of European Funds. Most start-ups financed from European funds through RSUP are in the Center region, numbering 1562 [73].

On average, the age of successful entrepreneurs is closer to 40 years than 20 years, and this is confirmed from an analysis carried out by the National Council of Small and Medium Private Enterprises in Romania, which concluded that most of the beneficiaries of these entrepreneurship non-refundable financing schemes are not very young entrepreneurs that are around 31 to 45 years and have an entrepreneurship education. Additionally, in the center region of Romania, many young entrepreneurs were beneficiaries of more than 1000 grants offered through SUNS and RSUP, so the authors conclude that entrepreneurship education was decisive in starting a new business [74].

We can conclude that the Romanian center region can be used for the 2018–2020 period as a relevant region for studying the education and the entrepreneurial intentions in the Romanian business market.

3. Methodology of Study

According to Zikmund, the survey is the best method available to a researcher when the objective of his research is to assess the opinion or perception of his respondents on issues of interest and at some points in time. This research aims to determine the influence of cultural values and entrepreneurial preparation on students' intention to engage in future entrepreneurial practices [75]. The research methodology used for this study was the questionnaire, and the longitudinal method applied involves the research of a group of young people in VET education and ends in a school year. The questionnaires measured the students' opinion of entrepreneurship courses offered by pre-university education and the students' intentions to engage in entrepreneurship in the nearest future.

For this study, the authors proposed the following hypotheses:

Hypothesis H1: *Pre-university entrepreneurial education positively influences the variables (attitudes and perceptions, motivations and aspirations, entrepreneurial activity) regarding the entrepreneurial intention of students.*

Hypothesis H2: *Entrepreneurial education studied in high school in any form positively influences students to become future entrepreneurs.*

Hypothesis H3: *Young people are influenced by variables (individual characteristics, entrepreneurial culture, funding sources) in the decision to start a business.*

The establishment of these research hypotheses required a collection of data based on questionnaires applied to young people in vocational and technical education in Romania and the statistical certification of the causal relationships included. The stages of this research study were as follows: data collection through the questionnaire, preparation and analysis of data, univariate analysis, testing of research hypotheses, interpretation of results, and setting up for conclusions. The scales used to measure constructions have been validated in previous studies. The preparation and statistical analysis of the data were carried out by elaborating the structure of the data matrix, encoding the answers to the questionnaire applied, entering the data into the SPSS application, and verifying the variables. The constructions were measured by 5-step Likert scales, and the data analysis sought to determine the proximity or distance of the distribution of the series of values from the Gaussian distribution by using histograms deviation generated by asymmetry (Skewness) and vaulting (Kurtosis).

The testing of the research hypotheses was done by using multiple linear regression in two steps. Each linear regression directly related to the research model was written in the form of the equation.

$$Y_i = (b_0 + b_1 \times 1_i + b_2 X_{2i} + \ldots \ldots b_n X_{ni}) + \varepsilon_i \tag{1}$$

In Equation (1), Y is the result variable (dependent variable); b_0 is the free term of the regression line, b_1 is the coefficient of the first predictor (first independent variable), $X_1 b_2$ is the coefficient of the second predictor (X_2), b_n is the n coefficient of the predictor, and ε_i represents the difference between the predicted and observed value of Y for the participant i. The aim will be to find the linear combination of independent variables that correlate the maximum with the dependent variable.

The assessment of the validity of the tested research hypotheses was made by analyzing the following elements: significance level (p-value), the non-standardized value of the regression function coefficient (β), determination coefficient (R^2), and the calculated value of the statistical test (t). The coefficient of determination (R^2) quantifies the proportion of the variation of the dependent variable that can be explained by the variation of the independent variable. The coefficient of determination (R^2) takes values between 0 and 1. If the value of the coefficient R^2 is close to 1 between the dependent and the independent variable, there is a linear, direct, and very strong connection. The t-test was used to test the regression model's parameters. If the significance level (p) is less than 0.05, β is significantly different from 0, and the relationship between the two variables is significant.

3.1. Population and Sample—Respondents

The implementation period of the first questionnaire was 90 days, between January and April 2019, with a duration of completion of the questionnaire between 5 and 7 min. The implementation period of the second questionnaire was 60 days between January and March 2020, with a completion time between 7 and 10 min. The target group that was the subject of the research carried out consists of young people and 11th and 12th graders from Romanian technological high schools, who are enrolled in services specialization because those specializations implement the practice firm method and implicitly have several hours of entrepreneurial education in the curricula. The practice firm method is a method of stimulating entrepreneurship among young people in the VET education system. The young respondents are between the ages of 15 and 20, and from the gender point of view, respondents are both female and male.

In this study, the authors set out to evaluate the importance of entrepreneurial education and entrepreneurial activities, as well as determine how the entrepreneurial ecosystem influences the entrepreneurial theme for the category of 15 to 20-year-old students in Romania, located in the Centre Region of this country, and included in the pre-university environment (the high school secondary education level). There is little empirical research performed on this topic in Romania, especially on entrepreneurship among young people. The authors started this investigation in the technological high schools of the Central Region, an area considered representative for the entire Romanian vocational and technical education system, based on the fact that the other seven Romanian development regions have the same curriculum in all technological schools for entrepreneurial education. The organizations in the Vocational Education and Technical System in Romania are coordinated by the National Centre for the Development of Vocational Education (CNDIPT), and the students investigated in the authors' study are students from the VET (Vocational Education Training) that are currently enrolled in the compulsory entrepreneurship courses offered by the field of the services. At the time of this study, the population of students in this category was composed of students from the Centre Development Region and were designated as the population of the high school and vocational education study ISCED 3 (International Standard Classification of Education) [76].

The National Statistical Institute of Romania states the total number of Romanian students for the 2018–2019 school year in high school education was 629,000 students. From the whole school population, the share of pupils enrolled in high school education was only 17.7%. Also, in terms of territorial distribution in Romania, 17.7% of all students were enrolled in secondary schools in the North-East region, 14.2% from the South-Muntenia region, 12.8% from the South-East region, 12.6% from the North-West region, 12.0% from the Bucharest-Ilfov region, 11.2% from the South-West Oltenia region, 11.5% from the Centre region, and 9.0% from the West region [77]. The analysis uses data obtained through a sociological survey carried out in the Central Region of Romania, which consists of six counties such as Alba, Brasov, Covasna, Harghita, Mureș, and Sibiu, and the authors have demonstrated that this region is a representative sample for the national level. The data collected during 2019 and 2020, through online research, was based on young people that have provided information on entrepreneurship education and entrepreneurial activities during high school learning. Based on this data, the authors started an investigation of entrepreneurship using the youth from the Central Region, which was considered representative of the Romanian education system in the field of vocational and technical education.

3.2. Instruments & Investigation Tools

The population investigated in the authors' study are students and teachers in VET education that are enrolled in the technological Romanian branch, who participate in the compulsory entrepreneurship courses offered by the services profile of this technological branch. At the time of this study, the population of students in this category was made up of students from the Central Development Region and was adopted as the population for the high school study from vocational education (ISCED level 3). The research method used for this study was represented by the questionnaires. In the case of the authors' study, the longitudinal method was applied, which involves the research of a group of young people from VET education and ends in a school year.

In Questionnaire 1, the student questionnaire on entrepreneurial culture and entrepreneurship in the vocational and technical education (VET) system was applied online using Google Forms. The respondents of this questionnaire were students in the 11th and 12th grades from the technological high schools in terms of services in the Central region of Romania. The authors received answers to questionnaires from 159 students enrolled in the 2018–2019 school year. In Questionnaire 2, the student questionnaire on entrepreneurship education in vocational and technical education units (VET) was applied online, using Google Forms, also to young people and students in the 11th and 12th grades from technological high school service profile in the Central region of Romania. The authors received an answer to the questionnaire from 254 students enrolled in the 2019/2020 school year. Finally, in Questionnaire 3, regarding teachers' opinions on entrepreneurship education in vocational and technical education units was applied online, using Google Forms, for teachers from technological high schools, teaching in vocational and technical education (VET) in the service profile from the Central region of Romania. In total, 59 teachers responded.

The questionnaires contain several parts: identifying elements on the topic of entrepreneurship, inputting elements on the purpose of this research, questions on the proposed theme, and socio-demographic questions to find out the respondent's profile. The questionnaires also include areas of interest: motivation to be an entrepreneur, experience in entrepreneurship, entrepreneurial culture, sources of funding, psychological traits, attitudes, behavior, and entrepreneurial intentions. All questionnaires received have been interpreted in SPSS (Statistical Package for the Social Sciences), a software for statistical data analysis, and have also been validated or invalidated or processed hypothesis with this software. The authors have also used the Microsoft Excel program from the Microsoft Office package for graphical analysis of the data results from the questionnaires applied to vocational and technical students from Romanian Development Region Center. The research was conducted online, and the respondents are students between the ages of 15 and 20 years.

3.3. Validity and limitations of the Study

The authors have demonstrated that these 472 respondents are representatives of the students and teachers involved in teaching activities in the technological branch, service profile high schools from the Romanian center region. They start their demonstration based on the data available from the National Institute of Statistics (NIS) regarding Romanian education [77].

The data from Table 1 demonstrate the importance of the center region in the Romanian education system, based on the regional distribution of students and teachers in different Romanian development regions.

Table 1. The number of enrolled students and teachers in high school by development regions, at the beginning of the 2018–2019 school year.

Romania Region	No. of Students	% of Students	No. of Teachers	% of Teachers
North-Est	111.666	18%	9.155	17%
South-Est	80.507	13%	6.282	12%
South-Muntenia	89.394	14%	6.986	13%
South-Vest-Oltenia	70.353	11%	5.882	11%
West	56.935	9%	5.232	10%
North-West	79.373	13%	8.213	15%
Center	66.191	11%	6.491	12%
București-Ilfov	75.336	12%	5.594	10%

Romanian high school education includes the following channels and profiles: theoretical branch (humanistic and real profiles) that represent 50%, technological chain (technical profiles, services, natural resources, and environmental protection) that represent 50%, and vocational chain (military, theological, sports, artistic, and pedagogical profiles) that represent 50%. These courses are meant to help the student in choosing his future career according to affinities.

The technological chain used in our study contains three profiles. The technic profile represents 43% of the total technological chain, the natural resource profile represents 16%, and our studied profile from the technological branch represents 41% from the whole technological branch on a national scale.

Based on these two previous percent rate and also on the national student and teacher distribution, the authors assume that the percent of service profile from the technological branch represent 17% of the total number of high school students, and also based on the regional distribution, the number of students and teachers from this profile in the center region is 11.044 respective 767.

The conclusion is that the 59 teachers that answered the third survey represent 8% of the total number of technological service teachers from this region (767), and the 472 students that answer to the first and second survey represent 4% of the total number of technological service students from this region (11.044), and both percentages are representative of the population for the results of the study.

Table 2 present the respondent characteristics of the 472 students that answer to the first and second survey.

Table 2. The number of responses corresponding to the high school service profiles in the Questionnaires.

Respondents' Characteristics	Absolute Frequencies (N) 2020 Respondents	Absolute Frequencies (%)	Absolute Frequencies (N) 2019 Respondents	Absolute Frequencies (%)
Gender				
Male	99	39	46	71
Female	155	61	113	28
Service field of study (VET)				
Commerce	59	23	44	28
Economic	115	46	53	33
Tourism	56	22	43	27
Other profile	24	9	19	12
TOTAL N	254		159	

3.4. Reliability of the Instrument

After receiving data from the respondents, these data were processed accordingly using descriptive statistics, such as media, module, median, and standard deviation, and they were used in the processing section of the questionnaires, as well as inferential statistics such as Cronbach's alpha coefficient to assess reliability, Pearson r moment correlation, and multiple linear regression. These statistical analysis tools were used to process the questionnaires applied to the target group. To transform the information from the questionnaires applied, the authors used the variables in SPSS: nominal, ordinal variables that are qualitative variables, and the range and ratio variables are of type quantities. To assess the reliability, the Cronbach coefficient was used. As indicated by Sekaran, a Cronbach's alpha coefficient of 0.70 and higher is considered reliable and acceptable [78]. Then, the authors assessed the reliability using Cronbach's alpha coefficient for the variables in Table 3.

Table 3. Summary of processed cases of variables. Reliability Statistics.

Variable	Cronbach's Alpha	Cronbach's Alpha Based on Standardized Items
AP	0.890	0.891
MA	0.858	0.857
AA	0.893	0.892
IC	0.984	0.985
EC	0.981	0.982
SF	0.995	0.995

Note: AP—attitudes and perceptions, MA—motivations and aspirations, AA—entrepreneurial activities, IC—individual characteristics, EC—entrepreneurial culture, and SF—sources of financing.

The information in Table 3 shows that the Cronbach alpha value is 0.984, which indicates a high level of internal consistency for research with this specific sample of entrepreneurship, and the correlation matrix shows the strength of the association between variables. As can be seen in Table 3, Cronbach's alpha for all variables is well above the threshold of 0.70 recommended by Sekaran, and it can be deduced that the study meets the reliability hypothesis. The above results indicate that the four reflective constructs have sufficient reliability.

4. Results

The overall objective of the research is to develop a coherent methodological approach, accompanied by the foundation and application of pragmatic tools, for analyzing the entrepreneurship in the VET education system in Romania. Tools that could be extended to the entire education system in Romania. In order to achieve this objective, a coherent set of specific research objectives was detailed, respectively:

Objective 1 Determination of the factors that determine the entrepreneurial intentions and innovation of young people in the VET education system in Romania.

Objective 2 Identification of the entrepreneurial ecosystem in VET education (professional and technical education) in Romania.

Objective 3 Promote a modern approach to the concept of technological entrepreneurship linked to entrepreneurial skills and connected with today's social needs.

Objective 4 Research and improvement of entrepreneurial skills in the Romanian pre-university education system as well as technological entrepreneurship.

Objective 5 Conduct an analysis on entrepreneurial education in terms of status, skills, content, and values in order to advance proposals for improving entrepreneurial skills in Romanian pre-university education.

Objective 6 Determine young people's motivation to start a business between the ages of 16 and 20.

Objective 7 Facilitating the transition from school to the active life of students through the use of methods of entrepreneurship and technological entrepreneurship.

4.1. Data Description and Objectives Achievement

This section is based on Questionnaire 1, the student questionnaire on entrepreneurial culture and entrepreneurship in the vocational and technical education (VET) system and was applied online in 2019 using Google Forms with 159 respondents. Also, this section was used for supporting Objectives 4, 5, and 6.

An important question of the survey was focused on information showing the behavioral sides of young people on attitudes and perceptions in the study of entrepreneurship in the Romanian pre-university education, as can be seen in Table 4.

Table 4. Which aspects of behavior regarding attitudes and perceptions, do you think that can be developed in the study of entrepreneurship? (Question 1).

Variable	Absolute Frequencies (N = 154) Respondents 1	Absolute Frequencies (%)
1.1. It is easy to start a business	30	19%
1.2. I have entrepreneurial intentions	53	34%
1.3. I have a skills and knowledge	42	27%
1.4. I am afraid of failure	22	15%
1.5. Other	7	5%

Students considered that many perceptions can be developed by studying entrepreneurship in the pre-university education system, so in Table 4, it can be observed that 34% of them have entrepreneurial intentions and 27% of them have acquired skills and knowledge of entrepreneurship. However, 15% of them felt they had a fear of failure and the other 5% consider that they have other behavioral sides developed.

Another question about the students' behavior is the motivations and aspirations of young people for entrepreneurship, as can be seen in Table 5.

Table 5. What are the motivations and aspirations for entrepreneurship during high school (Question 2).

Variable	Absolute Frequencies (N = 154) Respondents	Absolute Frequencies (%)

Students also believe that motivations and aspirations help them to better manage the company (36%), but also to start a business in terms of innovation and technology for 23% of them. However, there are few who would like to take control, so 13% prefer to continue the family business.

Another question of the study is focused on entrepreneurial activity. Obviously from the previous responses, the results are aimed at the respondent's opinion on entrepreneurship and technology, and the results show that the entrepreneurial idea is a necessity for economic growth. Their role in technology-oriented businesses can be a way to start home-based businesses while maintaining a perfect family-life balance, as can be seen in Table 6.

Table 6. What are the reasons for participating in entrepreneurial activities during high school? (Question 3).

Variable	Absolute Frequencies (N = 154) Respondents	Absolute Frequencies (%)
3.1. An entrepreneurial activity as an employee	56	36%
3.2. An entrepreneurial activity as an employer	41	27%
3.3. An entrepreneurial activity on technology and innovation	44	29%
3.4. Other	13	9%

The results gathered show 36% of respondents said they want entrepreneurial activity as an employee, 27% want entrepreneurial activity as an employer, and 29% want an entrepreneurial activity on technology and innovation. However, there are 8% of them who want something else. Based on the above results, it is recommended that entrepreneurship should become a compulsory discipline and must be taught in all profiles of pre-university education, not only in the service profiles. In other European countries, this is already a common practice, and it should be also adopted in our country. As for the profile of the company that students would like to set up, technology and trade are at the top of the list—options that are in line with market opportunities. Interestingly, 20% of students choose companies with a profile that can bring innovation to the market, while previously, 74% of students said they like to work and innovate.

It can be concluded that young people develop a set of social, emotional, cognitive, behavioral, and entrepreneurial skills needed to lead a balanced life. Students will be empowered, which will give them self-confidence. They will be able to think freely and creatively and innovate. Last but not least, they will have the ability to find their own appropriate solutions for the efficient management of the challenges in everyday life.

Table 7 shows that in the Romanian pre-university education system, respondents consider that skills should be improved the most (53%), then capacity skills (27%), and finally, knowledge (20%). The European Commission also stressed that knowledge, skills, and attitudes are essential and must be linked to the competence of an entrepreneurial spirit.

Table 7. Which of the following skills needs to be improved on entrepreneurship in schools? (Question 4).

Variable	Absolute Frequencies (N = 154) Respondents	Absolute Frequencies (%)
4.1. Knowledge	31	20%
4.2. Capacity skills	41	27%
4.3. Skills	82	53%

Individual skills, knowledge, and aspirations greatly influence the success of an entrepreneur. These data show that students are aware of their own qualities that can provide them with a career in entrepreneurship. However, 14% of students admit that they do not like to work or innovate, associating these qualities with a possible career in entrepreneurship. In this case, it can be assumed that these students want other professions in the future.

4.2. Descriptive Statistics for Testing Hypotheses and Other Objectives Achievement

This section is based on the second questionnaire, which was the student questionnaire on entrepreneurship education in vocational and technical education units (VET) system and was applied online in 2020 using Google Forms, and we received 254 responses. Also, this section was used for supporting Objectives 1, 2, and 3.

According to Zikmund (2005), the descriptive procedure is useful for obtaining summary comparisons of approximate normally distributed scale variables and for easily identifying unusual cases between those variables. To give a descriptive perspective of the study constructs, a descriptive analysis was performed, and the tables containing the results are shown below in Tables 8–10.

Table 8. Descriptive statistics of individual characteristics.

Variable	N Statistic	Minimum Statistic	Maximum Statistic	Mean	Std. Deviation Statistic
Q.1.1	254	1	5	4.00	1.13
Q.1.2	254	1	5	3.59	1.24
Q.1.3	254	1	5	4.18	1.11
Q.1.4	254	1	5	4.24	1.06
Q.1.5	254	1	5	3.87	1.20
Valid N	254				

Mean Average: 3.96. Note: Individual characteristics–Q.1.1—I like to be independent, Q.1.2—I like to be a leader, Q.1.3.—I take responsibility for what I do, Q.1.4—I take decisions and risks for what I do, and Q.1.5.—I like to work and innovate.

Table 9. Descriptive statistics of entrepreneurial culture.

Variable	N Statistic	Minimum Statistic	Maximum Statistic	Mean	Std. Deviation Statistic
Q.2.1	254	1	5	2.70	1.35
Q.2.2	254	1	5	2.29	1.33
Q.2.3	254	1	5	2.05	1.47
Q.2.4	254	1	5	2.92	1.43
Q.2.5	254	1	5	2.44	1.47
Valid N	254				

Mean Average: 2.48. Note: Entrepreneurial culture—Q.2.1. I have specific skills and competencies in entrepreneurship, Q.2.2. I have participated in courses or workshops on entrepreneurship, Q.2.3. I participated in the courses offered by Junior Achievement., Q.2.4. I followed the Entrepreneurial Education-class X module, and Q.2.5. I participated in entrepreneurial events (fairs, workshops).

Table 8 presents the descriptive statistics indicating that for the five items on individual characteristics perceived among respondents, the average is between 3.87 for the innovation item and 4.24 for the decision-making and risk item. The average for psychological traits is 3.96, which indicates that respondents are moderate in their perception of entrepreneurship.

Table 10. Descriptive statistics of sources of financing for setting up a company.

Variable	N Statistic	Minimum Statistic	Maximum Statistic	Mean	Std. Deviation Statistic
Q.3.1	254	1	5	2.08	1.42
Q.3.2	254	1	5	2.16	1.48
Q.3.3	254	1	5	2.05	1.40
Q.3.4	254	1	5	2.16	1.43
Q.3.5	254	1	5	1.94	1.31
Valid N	254				

Mean Average: 2.08. Note: Sources of financing—Q.3.1 Accessing the Start-Up program, Q.3.2. Access to other European funds, Q.3.3. Financial leasing, Q.3.4. Bank credit, and Q.3.5. Business incubators.

In Table 9, the authors analyzed the average for the five elements that constitute the entrepreneurial culture and found this indicator between 2.05 for extra-curricular entrepreneurial education and 2.72 for entrepreneurial activity skills and abilities. The average of the 5 items is 2.48, which indicates that respondents are moderate about the entrepreneurial culture.

Table 10 presents the average score for items of funding (2.08), which shows that these respondents are moderate in terms of sources of funding for a new firm in the market. Since the average is between 1.94 for business incubators and 2.16 for Start-Up programs for accessing European funds, these very small averages indicate that respondents do not have enough information to attract funds for a future company.

In Table 11, the authors analyzed a correlation between independent variables with the dependent variable on psychological traits, entrepreneurial culture, and funding sources.

Table 11. Correlation matrix of major variables.

Variable	Indicator	Q1.1	Q1.5	Q2.1	Q2.2	Q2.4	Q3.1	Q3.4
IC	Q.1.1	1						
N = 254	Q.1.5	0.962	1					
EC	Q.2.1	0.792	0.804	1				
N = 254	Q.2.2	0.731	0.732	0.922	1			
	Q.2.4	0.846	0.860	0.931	0.918	1		
SF	Q.3.1	0.651	0.664	0.868	0.943	0.861	1	
N = 254	Q.3.4.	0.674	0.692	0.879	0.944	0.870	0.876	1

Correlation is significant at the 0.01 level (2-tailed). Note: IC—Individual characteristics, EC—entrepreneurial culture, and SF—Sources of financing.

Pearson product-moment correlation analysis was used to determine the nature, direct or inverse, and the degree of association between variables, while multiple regions were used to determine the explanatory power of independent variables over the dependent variable. This is a common measure of the relationships between numerical variables measured on the Likert scale. The correlation matrix shows the strength of the association between the variables and is presented above in Table 11. The table above shows the cross-correlation coefficients of the main constructs in this study. As the table indicates, the entrepreneurial intention is significantly and positively related to adequate perception with the correlation coefficient ($p < 0.01$). Therefore, the authors conclude that there is a strong, significant, and positive association between these two constructions.

4.3. Statistical Analyzes and Study Results

The authors' further attempt was to validate the two hypotheses proposed at the beginning of the study by using the statistical techniques provided by SPSS as well as from the above graphical analysis where the authors have interpreted the questionnaires applied in 2019–2020. The purpose of analyzing data and testing the hypotheses was aimed at establishing the approximation or distance of the distribution of the values from the

Gaussian distribution by using histograms and the deviations generated by asymmetry (Skewness) and vaulting (Kurtosis).

To demonstrate Hypothesis H1, the authors started by quantifying the degree of implementation of variables (attitudes and perceptions, motivations and aspirations, entrepreneurial activity) from the questionnaires applied. Analyzing the independent variable with the dependent variable by calculating its variable correlation matrix, the authors have observed that there is a positive relationship between independent variables and the dependent variable, which explains that an increase in an independent variable has an increasing effect on the other.

The link between variables is highlighted in Table 12 by the multiple correlation coefficient obtained in this analysis (R = 0.969)—a value that shows a very good link between the dependent variable and the independent variables. The model is valid because F has a high level of significance and indicates a very significant relationship between the established variables. The model explains 96% of the total variation of the dependent variable (R^2 = 0.938), and the remaining 4% is influenced by other factors. Data on the linear multiple regression analysis is required to test Hypothesis H1, and it can be observed that R^2 has adequate values on students' intentions to engage in future entrepreneurial efforts. In other words, the intention for future entrepreneurial activities can be promoted by the dissemination of information that will encourage students to combine their attitude with a career in entrepreneurship and an acceptable and appropriate career after graduation. Based on the results obtained from statistical analyses, the authors conclude that the Hypothesis H1 is accepted, and it is a correlation between the pre-university entrepreneurial leading and positively influences the variables (attitudes and perceptions, motivations and aspirations, entrepreneurial activity) regarding the entrepreneurial intention of the students.

Table 12. Multiple regression analysis between the independent variables and the dependent variable on entrepreneurial intentions.

Model Summary									
Model	R	R Square	Adjusted R Square	Std. Error of the Estimate	Change Statistics				
					R Square Change	F Change	df1	df2	It's getting you out of here F Change
1	0.969 [a]	0.939	0.935	0.121	0.939	220.231	10	143	<0.001

ANOVA [b]						
Model		Sum of Squares	df	Mean Square	F	It's getting you out of here
1	Regression	32.342	10	3.234	220.231	<0.001 [c]
	Residual	2.100	143	0.015		
	Total	34.442	153			

[a] Predictors: (Constant) An entrepreneurial activity on technology and innovation. I would continue my family business. An entrepreneurial activity as an employee, I would like to earn a living, fear of failure, an entrepreneurial activity as an employer, starting a company, I would like to start a business in the field of technology, I would like to start a business, skills, and knowledge. [b] Dependent Variable: Entrepreneurial intentions. [c] Predictors: (Constant) An entrepreneurial activity on technology and innovation. I would continue my family business. An entrepreneurial activity as an employee, I would like to earn a living, fear of failure, an entrepreneurial activity as an employer, starting a company, I would like to start a business in the field of technology, I would like to start a business, skills, and knowledge.

To demonstrate Hypothesis H2, the authors started by quantifying the degree of implementation of high school entrepreneurial education accumulated during the 4 years. In the questionnaires applied on the exercise firm and implicitly entrepreneurial education, the authors found what field students want to set up their own company. From the correlation matrix analyzed, the authors have observed that there is a positive relationship between independent variables and the dependent variable, which explains that an increase in an independent variable has an increasing effect on the other.

The link between variables is highlighted in Table 13 by the multiple correlation coefficient obtained in this analysis (R = 0.982)—a value that shows a very good link between the dependent variable and the independent variables. The model is valid because F has a high level of significance and indicates a very significant relationship between the established variables. The model explains 98% of the total variation of the dependent variable ($R^2 = 0.965$), and the remaining 2% is influenced by other factors.

Table 13. Multiple regression analysis between independent variables and the dependent variable on the exercise firm for 159 respondents.

Model Summary								
Model	R	R Square	Adjusted R Square	Std. Error of the Estimate	Change Statistics			
					F Change	df1	df2	It's getting you out of here F Change
1	0.982 [a]	0.965	0.964	0.250	1405.150	3	155	<0.001

ANOVA [b]						
Model		Sum of Squares	df	Mean Square	F	It's getting you out of here
1	Regression	263.409	3	87.803	1405.150	<0.001 [c]
	Residual	9.685	155	0.062		
	Total	273.094	158			

[a] Predictors: (Constant) The exercise company helped you develop your personal skills to be a future entrepreneur, the exercise company helped you develop your entrepreneurial behavior, and the exercise company helped you develop to be an entrepreneur. [b] Dependent Variable: The exercise company helped you develop your intention to be a future entrepreneur. [c] Predictors: (Constant) The exercise company helped you to develop your personal skills to be a future entrepreneur, the exercise company helped you to develop an entrepreneurial behavior, and the exercise company helped you to develop to be an entrepreneur.

The analyses made over the entrepreneurial intentions show a strong link between variables that is highlighted in Table 14 by the multiple correlation coefficient obtained in this analysis (R = 0.969)—a value that shows a very good link between the dependent variable and the independent variables. The model is valid because F has a high level of significance and indicates a very significant relationship between the established variables. The model explains 96% of the total variation of the dependent variable ($R^2 = 0.939$), and the remaining 4% is influenced by other factors.

Table 14. Multiple regression analysis between the independent variables and the dependent variable on entrepreneurial intentions for 254 respondents.

Model Summary								
Model	R	R Square	Adjusted R Square	Std. Error of the Estimate	Change Statistics			
					F Change	df1	df2	It's getting you out of here F Change
1	0.969 [a]	0.939	0.935	0.121	220.231	3	254	<0.001 [b]

[a] Predictors: (Constant) Entrepreneurial activity in the exercise company helped you develop entrepreneurial behavior, the exercise company helped you develop your personal skills to be a future entrepreneur. [b] Predictors: (Constant) Entrepreneurial activity in the exercise company helped you develop entrepreneurial behavior, the exercise company helped you develop your personal skills to be a future entrepreneur.

In conclusion, the model achieved a good precision of entrepreneurial intent among the students in this study. Finally, from the analysis of the data and the interpretation of the above data, they indicate a significant relationship between the independent variable and the dependent variable on the other side. This validates Hypothesis H2, where the entrepreneurial activity in the exercise firm positively influences them to become future entrepreneurs since they already have contact with a virtual firm in which all the activities of a firm are simulated.

The link between the variables is highlighted in Table 15 by the multiple correlation coefficient obtained in this analysis (R = 0.949)—a value that shows a very good link between the dependent variable and the independent variables. The model is valid because F has a high level of significance and indicates a very significant relationship between the established variables.

Table 15. Multiple regression analysis between the independent variables and the dependent variable.

Model Summary								
Model	R	R Square	Adjusted R Square	Std. Error of the Estimate	Change Statistics			
					F Change	df1	df2	It's getting you out of here F Change
1	0.974 [a]	0.948	0.947	0.278	644.525	7	246	<0.001
1	0.949 [b]	0.900	0.897	0.43534	369.654	6	247	<0.001

[a] Predictors (constant) IC—Individual characteristics, EC—entrepreneurial culture, SF—Sources of financing. [b] Predictors (constant) EC—entrepreneurial culture, SF—Sources of financing, and IC—Individual characteristics.

The model explains 94% of the total variation of the dependent variable ($R2 = 0.900$), and the remaining 6% is influenced by factors. The data on the linear multiple regression analysis is needed to test the hypothesis in this thesis, and it can be seen that R2 has adequate values regarding students' intentions to engage in future entrepreneurial efforts.

The model is valid because F has a high level of significance and indicates a very significant relationship between the established variables.

The model explains 97% of the total variation of the dependent variable ($R2 = 0.948$), and the remaining 3% is influenced by other factors. The data on the linear multiple regression analysis is needed to test the hypothesis in this thesis, and it can be seen that R2 has adequate values regarding the intentions of students to engage in future entrepreneurial efforts.

After analyzing the above data and statistics, the author showed that Hypothesis H3, which says that students/young people are influenced by the variables (individual characteristics, entrepreneurial culture, funding sources) in the decision to start a business, is accepted because there are stable relationships between dependent variables and independent variables.

At the end of this section, the authors conclude that objectives 6 and 7 are supported, and also the three proposed hypotheses have been demonstrated and accepted.

5. Discussion

Entrepreneurship is important for job creation, economic growth, and innovation. It can also strengthen social inclusion and address societal challenges through policies of entrepreneurship, social entrepreneurship, and technological entrepreneurship. Innovation and entrepreneurship are in an interdependent relationship. Innovation can be considered a resource available to any entrepreneur, providing him with valuable tools for integration into the competitive business environment, adaptation and, of course, growth and development in a competitive business environment. Entrepreneurship supports innovation, as demonstrated by the policies of the European Union and taken over by the entrepreneurial policies in Romania. This study highlights that the new generations of Romanian entrepreneurs with unlimited access to information will change Romania's status as a "modest innovator," as mentioned in the European Innovation Scoreboard.

Economic growth is strongly correlated with the abundance of small, entrepreneurial firms; this relationship shows technological progress, stimulates product cycles, where growth is faster in the previous stages, and the importance of entrepreneurship. Instead, the evidence suggests that spatial differences in the fixed costs of entrepreneurship in the offer of entrepreneurs best explain the formation of clusters. Group learning experiences within an exercise firm are marked by the stimulation of critical thinking and the self-confidence that students gain during the implementation of this experimental learning strategy. It also contributes to the establishment of a relaxed working relationship between the educator and the educated. If at first students found it harder to organize, they got used to working in this way, and some of them are showing creativity. Creativity is an essential prerequisite for innovation, which is necessary for Romanian entrepreneurial organizations to become competitive in the European market. Vocational and technical education (VET) in Romania tries to train young students in entrepreneurial skills, skills, and attitudes by training and practicing them in a modern, flexible, and connected learning framework to the demands of the global business environment. From this perspective, an entrepreneurial innovation is being attempted, i.e., a transfer of technological entrepreneurship that is structured and most importantly sustainable.

The current results of the various studies show that entrepreneurship education encourages young people to start their own business and more, has a positive impact on the students' self-assessment and attitude towards entrepreneurship, as well as their aspirations and achievements. There have also been studies such as that of Rita R, Grazina S, and Daiva D, which have studied the factors that influence entrepreneurial intent, concluding that entrepreneurial intent is mostly influenced by personal factors, personality traits, which can be developed by following studies on entrepreneurship. Therefore, personality traits have been shown to have a direct impact on entrepreneurial intent, but this impact can become even more strengthened through entrepreneurial education [79].

Compared to other research made worldwide (28), especially in Romania (13) during the last 3 or 4 years, the authors tried to demonstrate that entrepreneurship education in high schools can influence the entrepreneurial intentions of the students and then the Romanian business market development.

In Romania, at the level of technological high schools, entrepreneurship should not be promoted and supported as an alternative for a professional career but should also be integrated into an ecosystem of entrepreneurial education at secondary, high school, and university level. Further, a national strategy with other European countries for a successful model could be developed. Internationally, entrepreneurship is a key element for job creation and the economic growth of a country. Supporting entrepreneurship has come to be a key priority and a solution to relaunch the economy. It follows from this research that young people certainly have concerns to develop businesses, a huge motivation to start their own projects. and a good perspective on entrepreneurial activities that are effective tools to increase their incomes. Their motivation is high, and they are also impressed by the newly arrived IT giants, considering that technology companies offer an economic boost both as an employer and as an employee.

A student's practice and activity in entrepreneurial activity in the practice firm should be considered professional experience for the student, and this is the opportunity to learn and obtain more skills and to identify which career or specialization best suits him or her socio-professional profile. In recent years, Romanian pre-university education tends to develop and capitalize on the integration of experimental learning strategies such as exercise firms, thus stimulating their idea of launching a commercial or social project. There are also situations where some students do not want to be actively involved in the work of exercise firms because they argue that they do not want to own their own business, and in this case, the teacher must shape the mentality of young people regarding entrepreneurship education: it is not only those who will launch their own business who have to work in an exercise firm. Entrepreneurial education forms and stimulates the entrepreneurial spirit required of any employee in a firm or institution, which depends very much, after all, on the efficient functioning of an institution or the development of the business.

In 2020, the EU education ministers met in Osnabrück to achieve a new strategy for VET education and achieve cross-border cooperation within the European Education Area, according to the Osnabrück Declaration (2020). The European strategy for 2020–2027 is built around three pillars: monitoring developments and assessing countries in need of modernizing education and training systems, providing policy advice to partner countries and EU institutions, and achieving a global knowledge hub for people and capital development in developing and transition countries [80].

In line with the identified trends and challenges, a greater focus should involve a change in workplace learning that offers specific key technical and entrepreneurial skills so as to facilitate the transition in the labor market. Learning and career development should also be improved by achieving learning paths with good integration and coordination between general education, as well as between formal, non-formal, and informal training, as well as for lifelong learning. Based on these and taking into account the Global Entrepreneurship Monitor reports [81], it is recommended that school authorities and relevant stakeholders engage in mass communication in the dissemination of information that favors perceived adequacy and perceived effectiveness, in addition to improving the quality of entrepreneurial training facilities in schools.

Romania has been part of the EU since 2007, and in the field of entrepreneurship, it is not far behind other states, especially in businesses developed by young entrepreneurs. The central region could be considered representative of the Romanian field of technological entrepreneurship because the industrial specificity is pronounced in the counties of Brasov and Sibiu—counties with important industrial traditions. The authors based this study on the fact that the central region of Romania is the Romanian region that has developed the most start-up businesses based on entrepreneurship schemes financed by the Romanian government and EU, which demonstrates the entrepreneurship potential of the region.

6. Conclusions

In conclusion, the authors have revealed that supporting entrepreneurship has become a key priority and a solution to revive the economy. This research shows that young people certainly have concerns about developing business, a huge motivation to start their own projects, and a good perspective on entrepreneurial enterprises that are effective tools to increase their income. Their motivation is high, and they are also impressed by the recent IT giants, considering that technology companies offer an economic boost either from the position of employer versus employee.

The study revealed that the students were determined to work successfully in the exercise company and in entrepreneurial activities, and they worked with pleasure. The very good results did not take long to appear, and the training within the exercise company proved to be interdisciplinary, oriented towards an experimental strategy (action and problems), focused on students, and inspired by practice.

The process of creating technological entrepreneurship is largely conditional on the trainers' expertise and their ability to identify, study, and stimulate students in different contexts, case studies, tools, and learning activities that are adapted to the economic and social context in which we are located.

Another conclusion was that basing a business on technology, in fact, entrepreneurship based on computer-assisted systems and methodologies, can not only bring innovation but also guarantees more profitability, in contrast to fewer investments that would otherwise have been needed in traditional enterprises of the same level. So, it is time for investors to bring the small- and medium-sized business sector into perfect alignment with technology-based businesses. As loans are important for financing start-ups, it is mandatory for banks to make policies and procedures to enable potential businesses to comply with financing options. A student's practice and activity in entrepreneurial activity in the practice firm should be considered professional experience for the student, and this is the opportunity to learn and obtain more skills and to identify which career or specialization best suits him or her socio-professional profile. In recent years, the Romanian pre-university education tends to develop and capitalize on the integration of experimental learning strategies such as exercise firms, thus stimulating their idea of launching a commercial or social project. There are also situations where some students do not want to be actively involved in the work of exercise firms because they argue that they do not want to own their own business. In this case, the teacher must shape the mentality of young people regarding entrepreneurship education: it is not only those who will launch their own business who have to work in an exercise firm. Entrepreneurial education forms and stimulates the entrepreneurial spirit required of any employee in a firm or institution, which depends very much, after all, on the efficient functioning of an institution or the development of the business. In line with the identified trends and challenges, a greater focus should involve a change in the workplace learning offers specific to key technical skills and entrepreneurial skills so as to facilitate the transition in the labor market. Learning and career development should also be improved by achieving learning paths with good integration and coordination between general education, as well as between formal, non-formal and informal training, as well as for lifelong learning. Based on these, it is recommended that school authorities and relevant stakeholders engage in mass communication about the dissemination of information that favors perceived adequacy and perceived effectiveness, in addition to improving the quality of entrepreneurial training facilities in schools.

The study shows that there is a very high trend for the establishment of innovation-based businesses to cope with worsening economic scenarios, especially when there is a high interest and motivation among young people to start innovation-based businesses, technology because they represent the future of any country.

7. Policy Implications and Study Limitations

Regarding the limitations of this study, even though it reveals the points of view of a representative population of students and professors from the Romanian primary and secondary education system, the results are limited in terms of their level of generalizability because of cultural contexts and participants' particularities from the central region. Another limitation is related to the item construction that measured the level of entrepreneur culture implementation, but also the future research design could be improved. Self-reporting and self-evaluation can also be risk factors because of social desirability.

In the field of policy implications of this study results, the authors believe that we can expect greater awareness of the need to develop and establish models of progress for entrepreneurship and technology entrepreneurship, rather than continuing to seek an approach to entrepreneurship education. We can also hope for researchers to identify characteristics of entrepreneurship education at all levels of education to a higher degree of certainty and with stronger empirical evidence than was possible. Also, the authors hope in the future that teachers will have access to qualifications and other support materials that allow them to choose from a wide variety of pedagogical tools and methods, enabling them to quickly identify and improve a teaching style and a progress strategy appropriate to their own students, and for the beginning, it must be created a built-in value for all students.

Based on this study, their educational experience in business, and over 20 years of entrepreneurial experience, the authors conclude that it is necessary to expect a better understanding and learning through achievement in the future. This can be integrated into education at all levels and for most subjects in the curricular area. Tools, methods, and concepts must be contextualized through education, leading to useful materials to support teachers and students for tasks performed through close collaboration between experienced teachers and dedicated to all levels of education and researchers in entrepreneurship and education.

Author Contributions: A.F. carried out the research based on his own questionnaires on online entrepreneurship through the partnership agreements concluded with the technological high schools of the Central Region of Romania. E.E. and L.L.-D. established the theoretical background and carried out the analysis of the Romanian entrepreneurship education system. All authors have read and agreed to the published version of the manuscript.

Funding: This research received no external funding.

Institutional Review Board Statement: The study was conducted according to the guidelines of the Declaration of Helsinki.

Informed Consent Statement: Informed consent was obtained from all subjects involved in the study.

Data Availability Statement: The data presented in this study are available on request from the first author.

Conflicts of Interest: The authors declare no conflict of interest.

References

1. Brancu, L.; Munteanu, V.; Gligor, D. Study on Student's Motivations for Entrepreneurship in Romania. *Proc. Soc. Behav. Sci.* **2012**, *62*, 223–231. [CrossRef]
2. Zamfir, A.-M.; Lungu, E.-O.; Mocanu, C. Study of Entrepreneurship Behavior among Higher Education Graduates in 13 European Countries. *Theor. Appl. Econ.* **2013**, *18*, 588.
3. Dodescu, A.O.; Botezat, E.A.; Cohut, I.C.P.; Borma, A. Antecedents, Experiences and Entrepreneurial Intentions among Economics Students. In Proceedings of the 12th LUMEN International Scientific Conference Rethinking Social Action. Core Values in Practice RSACVP 2019, Asociatia LUMEN, Iasi, Romania, 15–17 May 2019.
4. Chigunda, F.; Schnurr, J.; Hames-Wilson, D. *Youth Entrepreneurship: Meeting the Key Policy Challenges*; Wolson College, Oxford University: Oxford, UK, 2002.
5. Dodescu, A.O.; Cohut, I.C.P. Youth Entrepreneurship and Role Models at Local Level. Case Study: Bihor County, Romania. *Rethink. Soc. Action Core Val. Prac.* **2018**, *5*, 117–134. [CrossRef]
6. Schoof, U. *Stimulating Youth Entrepreneurship: Barriers and Incentives to Enterprise Start-Ups by Young People*; International Labour Office: Geneva, Switzerland, 2006.
7. Kuratko, D.F.; Ireland, R.D.; Covin, J.G.; Hornsby, J.S. A model of middle level managers' entrepreneurial behavior. *Entrepreneur. Theory Pract.* **2005**, *29*, 699–716. [CrossRef]
8. Kuratko, D.F.; Hodgetts, R.M. *Entrepreneurship: Theory, Process*; Practice; South–Western College Publishers: Mason, OH, USA, 2004.
9. Ratten, V. Sport-based entrepreneurship: Towards a new theory of entrepreneurship and sport management. *Int. Entrep. Manag. J.* **2011**, *7*, 57–69. [CrossRef]
10. Fernandez-Guadaño, J.; Lopez-Millan, M.; Sarria-Pedroza, J. Cooperative Entrepreneurship Model for Sustainable Development. *Sustainability* **2020**, *12*, 5462. [CrossRef]
11. Busu, C.; Busu, M. Modeling the Circular Economy Processes at the EU Level Using an Evaluation Algorithm Based on Shannon Entropy. *Process.* **2018**, *6*, 225. [CrossRef]
12. Briciu, A.; Briciu, V.-A.; Kavoura, A. Evaluating How 'Smart' Brașov, Romania Can Be Virtually via a Mobile Application for Cultural Tourism. *Sustainability* **2020**, *12*, 5324. [CrossRef]
13. Bican, P.M.; Brem, A. Digital Business Model, Digital Transformation, Digital Entrepreneurship: Is There A Sustainable "Digital"? *Sustainability* **2020**, *12*, 5239. [CrossRef]
14. Schepers, J.; Voordeckers, W.; Steijvers, T.; Laveren, E. Long-Term Orientation as a Resource for Entrepreneurial Orientation in Private Family Firms: The Need for Participative Decision Making. *Sustainability* **2020**, *12*, 5334. [CrossRef]
15. Yamaguchi, C.K.; Stefenon, S.F.; Ramos, N.K.; Dos Santos, V.S.; Forbici, F.; Klaar, A.C.R.; Ferreira, F.C.S.; Cassol, A.; Marietto, M.L.; Yamaguchi, S.K.F.; et al. Young People's Perceptions about the Difficulties of Entrepreneurship and Developing Rural Properties in Family Agriculture. *Sustainability* **2020**, *12*, 8783. [CrossRef]
16. Carriles-Alberdi, M.; Lopez-Gutierrez, C.; Fernandez-Laviada, A. The Influence of the Ecosystem on the Motivation of Social Entrepreneurs. *Sustainability* **2021**, *13*, 922. [CrossRef]
17. Global Entrepreneurship Monitor, 2016/17. Available online: https://www.gemconsortium.org/report/gem-2016-2017-global-report (accessed on 17 January 2021).
18. Kim, M.; Lee, J.-H.; Roh, T.; Son, H. Social Entrepreneurship Education as an Innovation Hub for Building an Entrepreneurial Ecosystem: The Case of the KAIST Social Entrepreneurship MBA Program. *Sustainability* **2020**, *12*, 9736. [CrossRef]
19. Isenberg, D. *The Entrepreneurship Ecosystem Strategy as a New Paradigm for Economic Policy: Principles for Cultivating Entrepreneurship*; Babson Entrepreneurship Ecosystem Project; Babson College: Babson Park, MA, USA, 2011.
20. Entrepreneurial Ecosystem Diagnostic Toolkit. Aspen Network of Development Entrepreneurs. 2013. Available online: https://www.aspeninstitute.org/wpcontent/uploads/files/content/docs/pubs/FINAL%2N0Ecosystem%20Toolkit%20 Draft_print%20version.pdf (accessed on 10 January 2021).
21. Isenberg, D. Introducing the Entrepreneurship Ecosystem: Four Defining Characteristics. Available online: https://www.forbes.com/sites/danisenberg/2011/05/25/introducing-the-entrepreneurship-ecosystem-four-definingcharacteristics/?sh=5bd1ada65fe8N (accessed on 12 January 2021).
22. Brătucu, G.; Lixăndroiu, R.C.; Constantin, C.P.; Tecău, A.S.; Chițu, I.B.; Trifan, A. Entrepreneurial University: Catalyst for Regional Sustainable Development. *Sustainability* **2020**, *12*, 4151. [CrossRef]
23. Andrez, P.; Ttaj, D.; Dalle, J.-M.; Romanainen, J. Specific Support Startups, Scaleups and Entrepreneurship in Romania, Horizon 2020 Policy Support Facility. (Finland, Expert. TWO: 10.2777/68810). Available online: https://rio.jrc.ec.europa.eu/library/specific-support-romania-final-report-start-ups-scale-ups-and-entrepreneurship-romania (accessed on 28 January 2021).
24. Bărbulescu, O.; Tecău, A.; Munteanu, D.; Constantin, C. Innovation of Startups, the Key to Unlocking Post-Crisis Sustainable Growth in Romanian Entrepreneurial Ecosystem. *Sustainability* **2021**, *13*, 671. [CrossRef]
25. Lewis, K.; Massey, C. *Youth Entrepreneurship and Government Policy*; New Zealand Centre for SME Research, Massey University: Palmerstone North, New Zealand, 2003.

26. Gibb, A.A. Stimulating New Business Development, What else besides EDP. In *Stimulating Entrepreneurship and New Business Development, Chapter 3*; ILO: Geneva, Switzerland, 1988.
27. OECD. *Putting The Young in Business: Policy Challenges for Youth N Entrepreneurship*; The LEED Programme, Territorial Development Division: Paris, France, 2001.
28. Henry, C.; Hill, F.; Leitch, C. Entrepreneurship education and training: Can entrepreneurship be taught? Part I. *Educ. Train.* **2005**, *47*, 98–111. [CrossRef]
29. Oncioiu, I.; Căpușneanu, S.; Constantin, D.-M.O.; Türkeș, M.C.; Topor, D.I.; Bîlcan, F.R.; Petrescu, A.G. Improving the Performance of Entities in the Mining Industry by Optimizing Green Business Processes and Emission Inventories. *Processes* **2019**, *7*, 543. [CrossRef]
30. Cárcel-Carrasco, J.; Gómez-Gómez, C. Qualitative Analysis of the Perception of Company Managers in Knowledge Management in the Maintenance Activity in the Era of Industry 4.0. *Processes* **2021**, *9*, 121. [CrossRef]
31. Ellis, K.; Williams, C. *Maximising Impact of Youth Entrepreneurship Support in Different Contexts, Back-Ground Report, Framework and Toolkit for Consultation*; Overseas Development Institute: London, UK, 2011.
32. García, A.I.; Álvarez, T.G.; Román, M.G.; Martín, V.G.; Merchán, D.T.; Zamudio, S.C. University Mentoring Programmes for Gifted High School Students: Satisfaction of Workshops. *Sustainability* **2020**, *12*, 5282. [CrossRef]
33. Parker, S.C. *The Economics of Self-Employment and Entrepreneurship*; Cambridge University Press (CUP): Cambridge, UK, 2004.
34. Cuervo, Á.; Ribeiro, D.; Roig, S. Entrepreneurship: Concepts, Theory and Perspective. Introduction. In *Entrepreneurship*; Cuervo, Á., Ribeiro, D., Roig, S., Eds.; Springer: Berlin/Heidelberg, Germany, 2007. [CrossRef]
35. Schumpeter, J. *The Theory of Economic Development: An Inquiry into Profits, Capital, Credit, Interest, and the Business Cycle*; Harvard University Press: Cambridge, UK, 1983.
36. Francis, D.H.; Banning, K. Who Wants to be an Entrepreneur? *J. Acad. Bus. Educ.* **2001**, *1*, 5–11.
37. Korunka, C.; Frank, H.; Lueger, M.; Mugler, J. The Entrepreneurial Personality in the Context of Resources, Environment, and the Startup Process—A Configurational Approach. *Entrep. Theory Pr.* **2003**, *28*, 23–42. [CrossRef]
38. Adekiya, A.A.; Ibrahim, F. Entrepreneurship intention among students. The antecedent role of culture and entrepre-neurship training and development. *Int. J. Manag. Educ.* **2016**, *14*, 116–132. [CrossRef]
39. García-Jurado, A.; Pérez-Barea, J.; Nova, R. A New Approach to Social Entrepreneurship: A Systematic Review and Meta-Analysis. *Sustainability* **2021**, *13*, 2754. [CrossRef]
40. Shapero, A.; Sokol, L. *The Social Dimensions of Entrepreneurship*; Urbana-Champaign's Academy for Entrepreneurial Leadership Historical Research Reference in Entrepreneurship: Champaign, IL, USA, 1982.
41. CEE (Consortium for Entrepreneurship Education). Entrepreneurship Education Everywhere. 2005. Available online: http://www.entre-ed.org/_entre/whitepaperfinal.pdf (accessed on 17 January 2021).
42. Drucker, P.F. *Innovation and Entrepreneurship*; Harper Collins Publishers: New York, NY, USA, 1993.
43. Lu, G.; Song, Y.; Pan, B. How University Entrepreneurship Support Affects College Students' Entrepreneurial Intentions: An Empirical Analysis from China. *Sustainability* **2021**, *13*, 3224. [CrossRef]
44. Butum, L.C.; Nicolescu, L.; Stan, S.O.; Găitănaru, A. Providing Sustainable Knowledge for the Young Graduates of Economic and Social Sciences. Case Study: Comparative Analysis of Required Global Competences in Two Romanian Universities. *Sustainability* **2020**, *12*, 5364. [CrossRef]
45. Cioca, L.-I.; Bratu, M. Sustainability of Youth Careers in Romania—Study on the Correlation of Students' Personal Interests with the Selected University Field of Study. *Sustainability* **2020**, *13*, 229. [CrossRef]
46. Frolova, Y.; Alwaely, S.; Nikishina, O. Knowledge Management in Entrepreneurship Education as the Basis for Creative Business Development. *Sustainability* **2021**, *13*, 1167. [CrossRef]
47. Yuan, C.-H.; Wang, D.; Mao, C.; Wu, F. An Empirical Comparison of Graduate Entrepreneurs and Graduate Employees Based on Graduate Entrepreneurship Education and Career Development. *Sustainability* **2020**, *12*, 10563. [CrossRef]
48. Edelhauser, E.; Ionică, A. A Business Intelligence Software Made in Romania, a Solution for Romanian Companies During the Economic Crisis, Computer Science and Information Systems; Journal ComSIS: Novi Sad, Serbia, 2014; Volume 11, pp. 809–823. Available online: http://www.comsis.org/archive.php?show=pprms-1304 (accessed on 20 January 2021). [CrossRef]
49. Hatos, A.; Hatos, R.; Bădulescu, A.; Bădulescu, D. Are Risk Attitudes and Individualism Predictors of Entrepreneurship in Romania? *Amfiteatru Economic* **2015**, *17*, 148–161.
50. Xavier, S.R.; Kelley, D.; Herrington, M.; Vorderwulbecke, A. Global Entrepreneurship Monitor (GEM) 2012 Global Report. 2013. Available online: https://www.gemconsortium.org/report/gem-2012-global-report (accessed on 17 January 2021).
51. Lupu-Dima, L.; Edelhauser, E.; Corbu, E.C.; Furdui, A. Innovative Method of Increasing the Quality of Management in Administration Using the Principles of Sharing Economy. *Qual. Access Success* **2019**, *20* (Suppl. 1), 507–512, ISSN 1582-2559.
52. Csaba, C.; Badulescu, A.; Cadar, O. Economic and Entrepreneurial Education in Romania in the European context. *WLC 2016 World LUMEN Congress. Logos Univers. Ment. Educ.* **2016**, 239–245. [CrossRef]
53. Antohi, V.-M.; Zlati, M.L.; Ionescu, R.V.; Neculita, M.; Rusu, R.; Constantin, A. Attracting European Funds in the Romanian Economy and Leverage Points for Securing Their Sustainable Management: A Critical Auditing Analysis. *Sustainability* **2020**, *12*, 5458. [CrossRef]
54. Păunescu, C.; Molnar, E. Country's Entrepreneurial Environment Predictors for Starting a New Venture—Evidence for Romania. *Sustainability* **2020**, *12*, 7794. [CrossRef]

55. Costache, C.; Dumitrascu, D.-D.; Maniu, I. Facilitators of and Barriers to Sustainable Development in Small and Medium-Sized Enterprises: A Descriptive Exploratory Study in Romania. *Sustainability* **2021**, *13*, 3213. [CrossRef]
56. Manning, L.; Smith, R.; Conley, G.; Halsey, L. Ecopreneurial Education and Support: Developing the Innovators of Today and Tomorrow. *Sustainability* **2020**, *12*, 9228. [CrossRef]
57. Hagebakken, G.; Reimers, C.; Solstad, E. Entrepreneurship Education as a Strategy to Build Regional Sustainability. *Sustainability* **2021**, *13*, 2529. [CrossRef]
58. Foss, N.J.; Klein, P.G.; Bylund, P.L. *Entrepreneurship and the Economics of the Firm*; SMG Working Paper No. 6/2011; Institut for Strategic Management and Globalization: Frederiksberg, Denmark. Available online: https://research-api.cbs.dk/ws/portalfiles/portal/58916605/SMG_WP_6_2011.pdf (accessed on 21 March 2021).
59. Vodă, A.; Haller, A.-P.; Anichiti, A.; Butnaru, G. Testing Entrepreneurial Intention Determinants in Post-Transition Economies. *Sustainability* **2020**, *12*, 10370. [CrossRef]
60. Ajzen, I.; Fishbein, M. *Understanding Attitudes and Predicting Social Behavior*; Prentice Hall: Englewood Cliffs, NJ, USA, 1980.
61. Martin, C. Needs and Perspectives of Entrepreneurship Education for Postgraduate Students. A Romanian Case Study. *J. Plus Educ. (JPE)* **2015**, *3*, 153–157.
62. Leovaridis, C.; Frunzaru, V.; Cismaru, D. Entrepreneurial education in romanian universities. In Proceedings of the 10th International Technology, Education and Development Conference, IATED Academy, Valencia, Spain, 7–9 March 2016; 2016; Volume 1, pp. 92–102.
63. Dan, M.C.; Popescu, A.I. Entrepreneurship Education in Romanian Universities: Developing Student Entrepreneurial Behaviour. In Proceedings of the 10th International Conference on Education and New Learning Technologies, Palma, Spain, 2–4 July 2018; pp. 10130–10138, ISBN 978-84-09-02709-5.
64. Furdui, A. Analysis of the Entrepreneurial Ecosystem from Romania, Annals of the University of Petroșani. 2019. Available online: https://www.upet.ro/annals/economics/pdf/2019/p1/Furdui.pdf (accessed on 21 January 2021).
65. Furdui, A.; Edelhauser, E.; Popa, E.I. Innovation Management Correlated with the Models of Development of Technological Entrepreneurship. *Qual. Access Success* **2019**, *20*, 513–518, ISMB 1582-2559.
66. Edelhauser, E.; Ionica, A.; Leba, M. Modern Management Using IT & C Technologies in Romanian Organizations. In *Transformations in Business & Economics*; Vilnius University: Vilnius, Lithuania, 2014; Volume 13, pp. 742–759. ISSN 1648–4460. Available online: http://www.transformations.khf.vu.lt/32b/article/moder (accessed on 19 January 2021).
67. Anser, M.K.; Khan, M.A.; Awan, U.; Batool, R.; Zaman, K.; Imran, M.; Sasmoko, S.; Indrianti, Y.; Khan, A.; Abu Bakar, Z. The Role of Technological Innovation in a Dynamic Model of the Environmental Supply Chain Curve: Evidence from a Panel of 102 Countries. *Processes* **2020**, *8*, 1033. [CrossRef]
68. Law on National Education No. 1/2011. Available online: http://legislatie.just.ro/Public/DetailsDocument/125150 (accessed on 17 January 2021).
69. Recommendation of the European Parliament and of the Council of 18 December 2006 on key competences for lifelong learning. Available online: https://eur-lex.europa.eu/legal-content/RO/ALL/?uri=CELEX%3A32006H0962 (accessed on 18 January 2021).
70. Yasir, N.; Mahmood, N.; Mehmood, H.; Babar, M.; Irfan, M.; Liren, A. Impact of Environmental, Social Values and the Consideration of Future Consequences for the Development of a Sustainable Entrepreneurial Intention. *Sustainability* **2021**, *13*, 2648. [CrossRef]
71. Densitate IMM Start-Up Nation 2020. Available online: https://activ-advisor.ro/wp-content/uploads/2020/05/Densitate-IMM-Start-up-Nation-2020.pdf (accessed on 20 March 2021).
72. Start-Up Nation 2018, Lista Rezultatelor Verificarilor in Cadrul Celei de-a Doua Editii a Schemei de Minimis. Available online: http://www.imm.gov.ro/ro/2019/06/03/start-up-nation-2018-lista-rezultatelor-verificarilor-in-cadrul-celei-de-a-doua-editii-a-schemei-de-minimis/ (accessed on 20 March 2021).
73. Romania Start-Up Plus si diaspora Start-Up Afla cate Afaceri Noi au fost Infiintate si in ce Domenii. Available online: https://www.fonduri-structurale.ro/stiri/22420/romania-start-up-plus-si-diaspora-start-up-afla-cate-afaceri-noi-au-fost-infiintate-si-in-ce-domenii (accessed on 20 March 2021).
74. Analiza CNIPMMR, Profilul Antreprenorului Beneficiar al Programului Start-Up Nation 2017. Available online: https://www.ceccarbusinessmagazine.ro/analiza-cnipmmr-profilul-antreprenorului-benefi-ciar-al-programului-start-up-nation-2017-varsta-intre-31-45-ani-studii-universitare-in-special-in-domeniul-economic-a4852/ (accessed on 20 March 2021).
75. Zikmund, W.G. *Sampling Designs and Sampling Procedures*; Business Research Methods: South Western, OH, USA, 2005.
76. Education System in Romania, School Year 2018-2019, Synthetic Data. Available online: https://insse.ro/cms/sites/default/files/field/publicatii/sistemuleducationalinromania_2018_2019_0.pdf (accessed on 17 December 2020).
77. National Institute of Statistics-Press Release 156 of June 25. 2019. Available online: https://insse.ro/cms/sites/default/files/com_presa/com_pdf/sistemul_educational_2019_r.pdf (accessed on 17 December 2020).
78. Sekaran, U. *Research Methods for Business*; Wiley and Sons: New York, NY, USA, 2008.
79. Remeikiene, R.; Startien, G.; Dumciuviene, D. Explaining Entrepreneurial Intention of University Students: The Role of Entrepreneurial Education. In Proceedings of the Management, Knowledge and Learning International Conference 2013, ToKnowPress, Zadar, Croatia, 19–21 June 2013.
80. Declaration from Osnabrück. Available online: https://epale.ec.europa.eu/ro/content/partile-interesate-din-ue-sunt-de-acord-asupra-educatiei-profesionale-ca-o-cale-de-urmat (accessed on 18 January 2021).

81. Dézsi-Benyovszk, A.; Nagy, Á.; Szabó, T.P. Entrepreneurship in Romania Country Report 2014. Available online: https://www.gemconsortium.org/report/49404 (accessed on 5 April 2021).

Article

Research Regarding the Energy Recovery from Municipal Solid Waste in Maramures County Using Incineration

Miorita Ungureanu [1], Juhasz Jozsef [1], Valeria Mirela Brezoczki [1], Peter Monka [2] and Nicolae Stelian Ungureanu [1,*]

[1] North University Centre at Baia Mare, Faculty of Engineering, Technical University of Cluj-Napoca, V. Babes St. 62, 430083 Baia Mare, Romania; miorita.ungureanu@cunbm.utcluj.ro (M.U.); jozsef.juhasz@cunbm.utcluj.ro (J.J.); valeria.brezoczki@cunbm.utcluj.ro (V.M.B.)

[2] Faculty of Production Technologies with the Seat in Prešov, Technical University of Kosice, Štúrova 31, 080 01 Prešov, Slovakia; peter.pavol.monka@tuke.sk

* Correspondence: nicolae.ungureanu@cunbm.utcluj.ro; Tel.: +40-745-298-976

Citation: Ungureanu, M.; Jozsef, J.; Brezoczki, V.M.; Monka, P.; Ungureanu, N.S. Research Regarding the Energy Recovery from Municipal Solid Waste in Maramures County Using Incineration. *Processes* **2021**, *9*, 514. https://doi.org/10.3390/pr9030514

Academic Editor: Elsayed Elbeshbishy

Received: 18 February 2021
Accepted: 10 March 2021
Published: 12 March 2021

Publisher's Note: MDPI stays neutral with regard to jurisdictional claims in published maps and institutional affiliations.

Copyright: © 2021 by the authors. Licensee MDPI, Basel, Switzerland. This article is an open access article distributed under the terms and conditions of the Creative Commons Attribution (CC BY) license (https://creativecommons.org/licenses/by/4.0/).

Abstract: This paper presents a part of the study referring to exploring Energy Recovery from Municipal Solid Waste in Maramures County. In order to analyze the possibility of energetic recovery of municipal solid waste (MSW), data referring to the management system of MSW from Maramures county were cumulated and processed in a first stage in order to estimate the quantity of municipal solid waste and its composition, which might be recovered energetically. In the next stage, samples of municipal solid waste were collected from landfills, which were submitted to specific processing and analyses. The experimental data were processed and in the end the energy potential of municipal solid waste from Maramures county was found. This study will help stakeholders and those involved in waste management to assess the possibility of energy recovery. The analysis of the study concluded that municipal solid waste in Maramures County is a potential source of renewable energy.

Keywords: municipal solid waste; calorific value; energy potential

1. Introduction

The sustainable management of the waste is essential for the society. In the prioritization of the waste management, the following hierarchy is known: Prevention, reuse, recycling, recovery, and disposal [1].

The management of a landfill is disadvantageous for the environment and involves covering some areas of ground, and that is why another essential sector, the energy sector, is considered a perfect alternative, as municipal solid waste (MSW) is classified as a source of energy [2].

The energy recovery from municipal solid waste might play an important role in the transition to a circular economy, on condition that the processes of prevention, reuse and recycling should be priority in the systems of waste management [3,4].

The solid municipal waste has an important calorific value and burning the waste in an incinerator may be used in order to generate electric energy or heat meant to heat the population [5].

The most common thermal treatment process for MSW is incineration (generally without being treated before), this method is considered to be the most reliable and economical form of energy recovery MSW [6,7].

The waste incineration using the last generation technology usually requires a minimum temperature of 850 °C for a dwelling period in the heating chamber of 2 s and a good turbulence, with the minimum content of oxygen specific for the systems (for example at least 3% excess of oxygen in the free gas after the incineration in a system with fluidized bed) [8].

An important advantage of waste incineration is the fact that it represents a rapid waste treatment method, with very large quantities being destroyed in a relatively short span of time. Modern incinerators reduce the volume of the waste by 95–96 percent, depending on the composition of the waste, on the degree of recovering certain materials such as metals, glass, and the recyclation degree [5]. The reduction of waste volume by incineration leads to the reduction of the area necessary for storing and using them in different purposes. Waste incineration permits the destruction of organic pollutants and substances [8]. Another advantage is the possibility to use the ashes and the solid waste in the road construction and cement industries [6].

The process of thermal treatment also reduces at zero the danger of infestation of the ground water by possible infiltrations of the leachate resulting in deposits, and reduces the methane emissions by the abolished landfills [4].

Another important advantage is the fact that the energy recovery leads to the reduction in the consumption of conventional fuels of heat or electricity.

Landfills have been shown to be a source of greenhouse gas (GHG) and carbon dioxide (CO_2) emissions [9,10]. In this context one advantage to be highlighted in the case of incineration is the total elimination of methane emissions and the quantity of CO_2 emissions of MSW incineration highly depends on the waste composition and plant technology [8]. Landfills also emit organic compounds and inorganic compounds that cause odors and health problems for local citizens [9]. A systemic approach of sustainable action is essential in ensuring a favorable housing climate for citizens, in maintaining habitats and protected areas [11].

A disadvantage of waste incineration is a series of emissions resulting from their combustion. But according to the literature, these emissions can be reduced so that they fall within the limits allowed by the laws and regulations on industrial emissions [12,13].

Combustion of MSW in the incinerator results in flue gases, especially CO, CO_2, H_2O, NOx, and, if applicable, SO_2. To a lesser extent, acid gases such as HCl and HF are also produced, and, last but not least, heavy metals and macromolecules with high stability and higher molecular weight (dioxins, Furan's and PCB's) [7]. Under certain conditions dioxins, furans, and similar gaseous components are only destroyed; the rate of organic molecules' destruction depends on the high temperature inside the furnace and the residence time of combustion gases in the incinerator [8].

The emissions resulting from MSW incineration are greatly reduced by combustion technologies. But apart from these measures, modern incinerators are equipped with filtering systems for the resulting emissions. So, the modern incinerator is an efficient combustion system, which produces energy and reduces waste to an inert residue with minimal pollution [14]. But the cost of an emission minimization technology for an incinerator can be up to 35% of the project cost [5].

To take into account the possibility of energy recovery by incineration, the amount of waste used for incineration must not be less than 50,000 tons per year [8], and the calorific value must be greater than 7 MJ/kg [15]. Energy efficiency for incineration facilities [16], life cycle energy assessment for incineration facilities, economic impact, and social impact assessment [17] are aspects that need to be taken into account.

Some problems can be encountered during management of solid wastes since they have a heterogeneous structure. For this reason, physical features of solid wastes, such as moisture content (MC), calorific value (heating) (HHV), and composition, should be well known for their management through suitable methods [11,18].

In order to evaluate the possibility of energy recovery of MSW by incineration, it is first necessary to analyze the waste generation method, and then the following characteristics of MSW: Composition, moisture content (MC), chemical characteristics, and calorific value (heating) [1,2].

In conclusion, we can highlight three advantages in the energy recovery from municipal solid waste: Environment protection, energy production from non-conventional sources, and hygiene reasons.

A systemic approach of sustainable action is essential in ensuring a favorable housing climate for citizens, in maintaining habitats and protected areas [18].

The main motivation of this research is to contribute to the reduction, and if possible, to the elimination of landfills and to obtaining energy, an essential element for the transition to a circular economy.

2. Materials and Methods

In the first part of our study, we will analyze waste generation in Maramures County, and later we will analyze the characteristics of municipal solid waste in terms of energy. In this context, our study begins from the premise that municipal solid waste is regarded as a chemical fuel and specific experiments and analyses are carried out in this context. A characterization of a chemical fuel involves. A characterization of a chemical fuel requires the knowledge, in addition to its physical state and origin, also of characteristics such as: Chemical composition, calorific value, and moisture [5]. Finally, the calculations performed in order to determine the energy potential of municipal solid waste in the county are presented.

2.1. Waste Generation in Maramures County

The amount of municipal solid waste and its composition undergo changes depending on the consumption habits of the population which are constantly changing [19]. The values and characteristics of municipal solid waste differ not only from one country to another, but also from one region to another, even from one neighborhood to another in the same city [20,21].

The research area, the Maramureș County, is located on the North of Romania. The chief town is the municipality of Baia Mare. The area of the county is of 6215 km^2. Maramureș County has more than 500,000 inhabitants. The county is made up of 2 municipalities: Baia Mare, Sighetul Marmatiei, 11 towns and 63 villages. Currently, all urban and rural localities in Maramureș County benefit from sanitation services.

Most of the waste collected in the county is disposed of by landfill in two landfills located in Satu Nou de Jos (Baia Mare) and Sighetu Marmației (Figure 1) with an area of about 22 ha and a capacity of 4.2 million cubic meters are managed by two private companies [22].

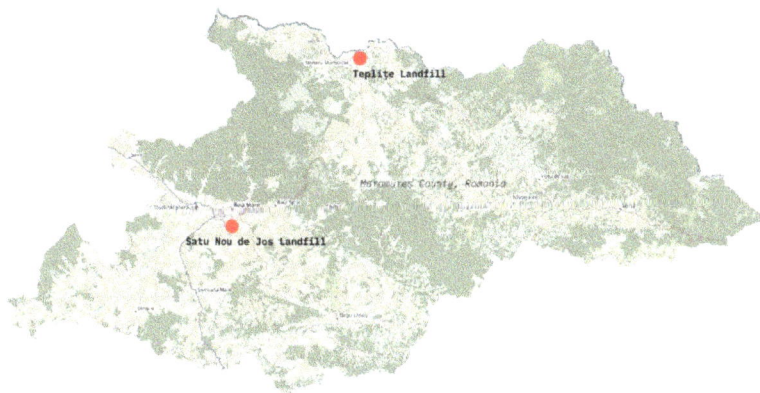

Figure 1. The landfills from Maramures County [23].

The total quantity of municipal solid waste generated in the year 2017 was of 86,382.3 tons of waste, and from this quantity 83,410 tons have been eliminated by removal [24]. In the year 2018 a quantity of municipal solid waste of 81,923.86 tons was generated, out of which 71,302 tons were deposited in landfills [25].

Our study considers the analysis of the energy potential of the amount of waste that is stored annually in the two landfills, in order to recover energy through incineration.

In order to determine the energy potential of municipal solid waste, we must know two main parameters: The calorific value of this waste and the amount of waste to be incinerated [2,26]. In addition to these two parameters, we have also determined other important characteristics in the study: The composition of the samples, the moisture of the samples and the elementary chemical composition of the samples.

2.2. The Estimated Municipal Solid Waste Quantity for Energy Recovery

In a first stage, based on the data provided by the accredited environmental agency in the county, the amount of waste available for incineration was determined, based on statistics from the last two years (2017 and 2018) and based on policies to increase recovery and recycling. Thus, based on the collected data, the total amount of municipal solid waste estimated for energy recovery is 53,325.87 tons per year and the distribution by waste categories is presented in Table 1. The amount available for incineration was obtained after deducting from the total amount of waste collected the part of recoverable and recyclable waste, glass, and inert waste from construction. This estimated annual amount of municipal solid waste for incineration and the chemical composition are shown in Table 1.

Table 1. Estimated annual amount of municipal solid waste (MSW) for incineration.

Physical Composition of MSW	Quantity (t)	Percentage
Paper, cardboard	5683.49	10.66
Plastic	5718.14	10.72
Wood	2571.99	4.82
Organic matter	30,794.78	57.75
Textile	3998.28	7.5
Bulk waste	4559.21	8.55
Total	53,325.87	100

The amount of waste generated per day to be incinerated is 146.1 t/day. In the case of incineration, facilities are justified above 100 t/day [6].

In conclusion, regarding the amount of waste generated per day, incineration is justified.

2.3. Collection and Preparation of Samples

Collecting waste samples for the study was the next important step. Obtaining a representative sample of one gram from a garbage truck full of waste is a very difficult operation, even if strict sampling and processing procedures are followed. Larger-scale instruments (1 kg) have been reported to determine the calorific value of waste, specifically designed for the analysis of municipal solid waste [5].

Samples were collected and prepared according to the study plan [27]. Due to the fact that the aim of the study is to determine the annual energy potential of municipal solid waste in Maramures County and to perform this calculation we refer to the total amount of waste presented in Section 2.2. The physical composition of samples was chosen to be consistent with the annual composition of waste. Under these conditions, based on the data collected regarding the percentage composition of MSW per year presented in Table 1, the samples were taken in compliance with the percentages. The municipal solid waste samples were collected from the two landfills in the county, Satu Nou de Jos (Figure 2a) and Teplița (Figure 2b). At the location of Satu Nou de Jos there is also a sorting station. The samples were taken from fresh waste arrived in the landfill, it was put in bags on the categories of materials presented in Table 1, the bags were transported to the EnyMSW laboratory, where the sorting and weighing by categories was performed.

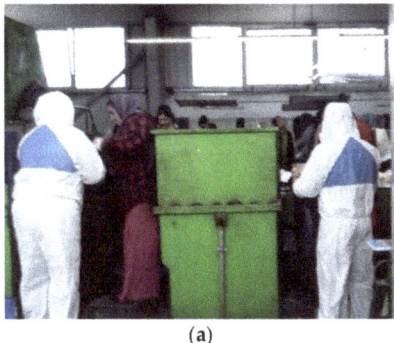 

(a) (b)

Figure 2. Samples collection: (**a**) From la Satu Nou de Jos; (**b**) from Teplite.

Each waste bag was marked with the date of collection, the place of collection, and the percentage composition (%) was classified into seven categories as follows:

- Organic matter in percentage of 57.75%;
- 10.72% plastic;
- 10.66% of paper and cardboard;
- 7.5% of textile materials;
- 4.82% of wood; and
- 8.5% bulk waste.

Thus, five samples, each 1 kg, were selected from the 2 landfills in Maramures County. The five samples were prepared and subjected to analysis and determination. After a first drying and the determination of the relative moisture, the samples were ground with Cutting Mill PULVERISETTE 15, the material resulting after grinding was sieved and homogenized (Figure 3).

Figure 3. Ground sample.

2.4. Moisture Content of the Municipal Solid Waste

The moisture content was performed by laboratory drying of an analytical sample, by drying the samples, until constant mass was reached. The results are presented in Table 2. The relative moisture was determined before grinding the samples by drying in the open air at the temperature of 20 °C with a relative moisture of the air of about 50%. The hygroscopic moisture was determined by drying in a laboratory oven at 105 °C.

Table 2. Moisture content of the MSW.

Sample Number	Relative Moisture Sample [%]	Hygroscopic Moisture [%]	Total Moisture [%]
Sample 1	3.69	40.01	42.22
Sample 2	2.18	42.08	43.34
Sample 3	3.97	41.21	43.54
Sample 4	2.13	41.93	43.17
Sample 5	2.78	40.84	42.48

The total moisture was calculated as follows [28]:

$$M_T = M_r + M_h(100 - M_r)/100 \quad (1)$$

where M_h is hygroscopic moisture [%] M_r and relative moisture [%].

2.5. Chemical Composition of Municipal Solid Waste

The chemical composition of municipal solid waste has most influence on the treatment method and recovery options [1,5]. In our study it is necessary to know the chemical composition of the samples to determine the net calorific value of the samples after obtaining experimentally a gross calorific value.

The chemical composition of municipal solid waste was determined by elementary chemical analysis in the Gas and Fuel Laboratory I.C.S.I. Rm. Vâlcea and resulted in the percentage content of carbon (C), hydrogen (H), sulfur (S), oxygen (O), and nitrogen (N) in the organic mass of the fuel for the samples. The chemical determination procedures for these elements are shown in Table 3 and the results of the determinations are shown in Table 4. The determinations were performed after the samples were dried.

Table 3. Chemical determination procedure.

Name of Determination	Procedure
Carbon	ASTM D 5373-16PS-AGC-15
Hydrogen	ASTM D 5373-16PS-AGC-15
Nitrogen	ASTM D 5373-16PS-AGC-15
Sulphur	PS-AGC-15
Oxygen	Calculus

Table 4. Chemical composition of the MSW sample.

Sample No	Carbon [%]	Hydrogen [%]	Nitrogen [%]	Sulphur [%]	Oxygen [%]
Sample 1	7.25 ± 1.07	3.74 ± 0.09	0.46 ± 0.02	S < LQ [1]	17.75
Sample 2	46.47 ± 1.33	6.24 ± 0.16	0.90 ± 0.03	0.50 ± 0.03	25.68
Sample 3	25.86 ± 0.74	2.96 ± 0.07	0.91 ± 0.03	0.30 ± 0.02	19.78
Sample 4	26.53 ± 0.76	2.98 ± 0.07	0.66 ± 0.02	0.29 ± 0.02	20.07
Sample 5	36.52 ± 1.05	4.97 ± 0.12	0.34 ± 0.01	S < LQ [1]	15.29

[1] LQ = 100 ppm.

3. Results

3.1. Determination of Calorific Value

From the 5 municipal solid waste representative samples, 5 small samples with a mass of 0.5 g were extracted and analyzed with the IKA C1/12 calorimeter (Figure 4) in accordance with the ISO 1928: 2009 standard [28]. The method for determining gross calorific values in the calorimeter IKA C1/12 is presented below. Thus, the mixed and homogenized samples were weighed with an accuracy of 0.0002 g.

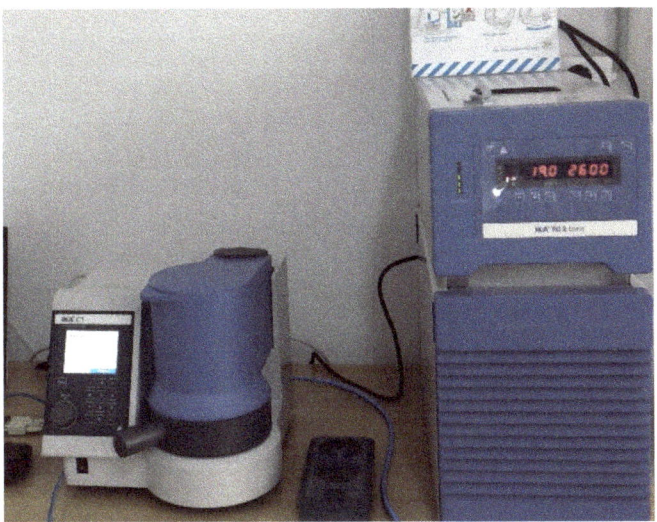

Figure 4. Calorimetric analysis for MSW sample.

The test sample was introduced into the decomposition vessel, where the burning under excess of oxygen in a closed container took place. The amount of heat resulting from this, measured by a previously calibrated system, allows the value of the calorific value of the sample to be determined.

The test sample is placed in a small bag, the bag in a ring-shaped crucible for burning and is closed in the decomposition vessel. A cotton thread, two electrodes and ignition wire are used for the ignition of the sample.

Five experiments were performed for each sample. The results of the experiments and of the determinations are presented in Table 5 and Figure 5.

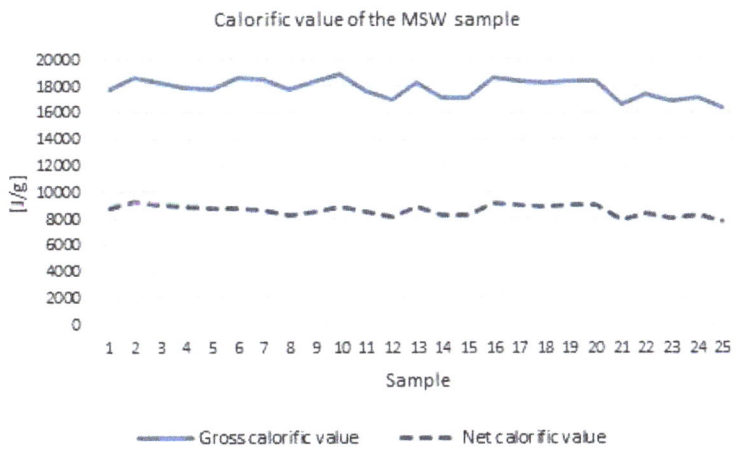

Figure 5. Variation of gross calorific value (GCV) and net calorific value (NCV).

Table 5. Calorific value of the MSW sample.

Sample	Gross Calorific Value (GCV) $(q_{V,gr,d})$ [J/g]	Net Calorific Value (NCV) $(q_{p,net,m})$ [J/g]
Sample 1	17,822	8800
	18,724	9321
	18,264	9055
	17,932	8863
	17,825	8801
Sample 2	18,633	8737
	18,532	8680
	17,828	8281
	18,350	8577
	18,958	8921
Sample 3	17,636	8530
	17,020	8182
	18,283	8895
	17,163	8263
	17,089	8221
Sample 4	18,643	9172
	18,437	9055
	18,254	8935
	18,390	9028
	18,389	9027
Sample 5	16,676	7941
	17,375	8343
	16,893	8066
	17,204	8245
	16,422	7795
Average	17,869.68	8629.36
Deviation	711.76	417.37
RSD (%)	3.98	4.84

The gross calorific value (GCV) resulted from the calorimeter analysis and the net calorific value (NCV) was calculated by introducing corrections that take into account moisture, hydrogen content of the sample, oxygen content of the sample, and nitrogen content of the sample.

In order to determine the net calorific value (NCV) the following calculus formula was used, according to ISO 1928:2009 [2,28]:

$$q_{p,net,m} = \{q_{V,gr,d} - 212 \times w_{H,d} - 0.8 \times [w_{O,d} + w_{N,d}]\} \times (1 - 0.01 \times M_T) - 24.43 \times M_T \quad (2)$$

where:
- $q_{p,net,m}$ is the net calorific value of the fuel at constant pressure and water content;
- $q_{V,gr,d}$ is the gross calorific value at constant volume and free water;
- $w_{H,d}$ is the hydrogen content of the sample, expressed as a percent mass fraction, of the moisture-free (dry) fuel;
- $w_{O,d}$ is the oxygen content of the sample, expressed as a percent mass fraction, of the moisture-free fuel;
- $w_{N,d}$ is the nitrogen content of the sample, expressed as a percent mass fraction, of the moisture-free fuel.
- M_T is the total moisture content, expressed as a percent mass fraction.

The resulted net calorific value (NCV) is of 8629.36 KJ/Kg.

3.2. Energy Recovery Potential of Municipal Solid Waste

The energy recovery potential of municipal solid waste per year can be calculated as follows [1,2,6,20]:

$$EP_{MSW/year} = W_{MSW/year} \cdot NCV/1000 \qquad (3)$$

where:
- $EP_{MSW/year}$ is the annual energy content of the treated waste, calculated on the basis of the lower net calorific value of the waste), [GJ];
- $W_{MSW/year}$-total waste quantity per year, [t];
- NCV—net calorific value, [kJ/kg].

Production of steam per year is [2]:

$$P_{steam/year} = W_{MSW/year} \times 2.5 \qquad (4)$$

If the option is for waste incineration plant with cogeneration of heat and electricity can achieve an optimum energy efficiency of some 80% (η_C) from total energy potential of the treated waste [8] and in this case the potential of recovered energy of MSW is:

$$EP_{CMSW} = \eta_C \times EP_{MSW} \qquad (5)$$

EP_{CMSW} is calculated in MWh.

Performing the calculations according to formulas (3)–(5) the following data result:
- Energy potential of the treated waste per year EP_{MSW} = 460,168 GJ = 127,824 MWh;
- production of steam per year = 133,314.7 t; and
- energy potential of electricity and thermal energy per year, $EP_{CMSW/year}$ = 102,259.58 MWh.

4. Discussion

The study carried out and presented in this paper is in line with the vision of the European Union which aims to introduce objectives based on the rejection of the linear economy in favor of the circular economy with the goal of recovering waste [4].

The study also aligns with Maramureș County's waste management policies and pursues the set objectives by both significantly reducing waste generation and increasing the recycling/recovery rate of waste [22].

As stated by the authors of the studies conducted for different locations, the energy potential of municipal solid waste depends on the calorific value of municipal solid waste and on the amount of municipal solid waste that is energy recovered. In addition to these parameters, the physical characteristics of the waste, the chemical composition and the moisture are also important parameters that influence the energy potential of MSW [1,2,6,11,14].

We noticed that the results obtained regarding the minimum quantity generated per year slightly exceeds the lower limit of 50,000 tons per year [8]. The MSW humidity, even though it is high, falls within the maximum limit of 50% [15], and the calorific value is higher than minimum limit of 7 MJ/kg [15].

After the development of the theoretical model and after the experimental researches, these two conditions: The minimum quantity and the humidity of municipal solid waste, represent in the opinion of the authors the main limitations related to the efficient application of the method.

This paper presents a first stage of the research regarding the Energy recovery of MSW by incineration, followed by other aspects related to the emissions resulting from incineration, to the heat treatment technologies of MSW with energy recovery and to the energy efficiency of the technologies to be approached in the future.

5. Conclusions

The energy potential of municipal solid waste depends on the calorific value of waste and on the amount of waste that is energy recovered. In addition to these parameters, the physical characteristics of the municipal solid waste, the chemical composition and the moisture are also important parameters that influence the heat treatment process. The analysis of the energy characteristics of municipal solid waste samples resulting in Maramures County indicates a satisfactory net calorific value (NCV): 8629.36 KJ/Kg and the moisture have the average 42.95%.

Municipal solid waste is currently generated at about 84,000 tons/year in Maramureș. From this quantity approximately 53,325.87 t/year could be recovered energetically. If chosen waste incineration plant with cogeneration of heat and electricity can achieve an optimum energy efficiency of some 80% from total energy potential of the treated waste and the energy recovery per year is 102,259.58 MWh.

These findings give reason to assume that the solid waste from Maramures County that is landfilled would have satisfactory energy characteristics to be recovered for energy by incineration, but with the recommendation of a feasibility study to continue this research on a much larger scale, which in addition to increasing the number of samples and tests to take into account the different periods of the year (for the humidity of the samples).

Author Contributions: Conceptualization, M.U.; methodology, M.U. and N.S.U.; validation, N.S.U., P.M.; formal analysis, M.U. and J.J.; investigation, J.J. and V.M.B.; resources, M.U.; data curation, M.U. and N.S.U.; writing—original draft preparation, M.U.; writing—review and editing, N.S.U. and P.M.; supervision, M.U.; project administration, M.U.; funding acquisition, M.U., P.M., J.J., and V.M.B. All authors have read and agreed to the published version of the manuscript.

Funding: This research was funded under the project "Energy Recovery from Municipal Solid Waste by Thermal Conversion Technologies in Cross-border Region" HUSKROUA/1702/6.1/0015, funded by Hungary—Slovakia—Romania—Ukraine ENI CBC Programme 2014–2020—of the European Union.

Institutional Review Board Statement: Not applicable.

Informed Consent Statement: Not applicable.

Data Availability Statement: For more details on the project behind this work, visit https://enymsw.eu/#/home (access on 18 January 2021).

Acknowledgments: The authors are grateful for the financial support of this project and also would like to thank to the stakeholders, especially to Maramureș County Council, Maramureș Environmental Protection Agency, National Environmental Guard—Maramureș County Commissariat and SC DRUSAL Baia Mare who, through their support, made it possible to carry out the experiments within the study.

Conflicts of Interest: The authors declare no conflict of interest. The funders had no role in the design of the study; in the collection, analyses, or interpretation of data; in the writing of the manuscript, or in the decision to publish the results.

References

1. Hussain, A.A.; Wesam, S.N.; Al-Rekabi, N.A.; Hamdan, A. Prediction of Potential Electrical Energy Generation from MSW of Basrah Government. In Proceedings of the 5th International Conference on Waste Management, Ecology and Biological Sciences (WMEBS-2017), Istanbul, Turkey, 15–18 May 2017; pp. 172–181.
2. Anshar, M.; Ani, F.N.; Kader, A.S. The energy potential of municipal solid waste for power generation in Indonesia. *J. Mek.* **2014**, *37*, 42–54.
3. Scarlat, N.; Fahl, F.; Dallemand, J.-F. Status and Opportunities for Energy Recovery from Municipal Solid Waste in Europe. *Waste Biomass Valorization* **2019**, *10*, 2425–2444. [CrossRef]
4. Rada, E.C.; Ragazzi, M.; Torretta, V.; Castagna, G.; Adami, L.; Cioca, L.I. Circular Economy and Waste to Energy. In Conference Proceedings, May, Technologies and Materials for Renewable Energy, Environment and Sustainability. *AIP Conf. Proc.* **2018**, *1968*, 030050-1–030050-6. [CrossRef]
5. Williams, T.P. *Waste Treatment and Disposal*, 2nd ed.; John Wiley & Sons Ltd.: Hoboken, NJ, USA, 2005; ISBN-13 978-0-470849132, ISBN-10 0470849134.

6. Alzate, S.; Restrepo-Cuestas, B.; Jaramillo-Duque, A. Municipal Solid Waste as a Source of Electric Power Generation in Colombia: A Techno-Economic Evaluation under Different Scenarios. *Resources* **2019**, *8*, 51. [CrossRef]
7. Puna, J.F.; Santos, M.T. Thermal Conversion Technologies for Solid Wastes: A New Way to Produce Sustainable Energy. *Waste Manag.* **2010**, 89–124. [CrossRef]
8. Grech, H.; Neubacher, F. *Waste-to-Energy in Austria*; Austrian Federal Ministry of Agriculture, Forestry, Environment and Water Management. Available online: https://www.iswa.org/fileadmin/galleries/Publications/White%20Book/Whitebook_20 Incineration_2009%5B1%5D.pdf (accessed on 15 May 2020).
9. Pecorini, I.; Rossi, E.; Iannelli, R. Mitigation of Methane, NMVOCs and Odor Emissions in Active and Passive Biofiltration Systems at Municipal Solid Waste Landfills. *Sustainability* **2020**, *12*, 3203. [CrossRef]
10. Zhang, C.; Xu, T.; Feng, H.; Chen, S. Greenhouse Gas Emissions from Landfills: A Review and Bibliometric Analysis. *Sustainability* **2019**, *11*, 2282. [CrossRef]
11. Vieru, D. NOMOGRAMA of a Landfill (msw)—Setting m Parameter Values. *Atmos. Clim. Sci.* **2017**, *7*, 436–454.
12. Lu, J.-W.; Zhang, S.; Hai, J.; Lei, M. Status and perspectives of municipal solid waste incineration in China: A comparison with developed regions. *Waste Manag.* **2017**, *69*, 170–186. [CrossRef] [PubMed]
13. Lee, U.; Han, J.; Wang, M. Evaluation of landfill gas emissions from municipal solid waste landfills for the life-cycle analysis of waste-to-energy pathways. *J. Clean. Prod.* **2017**, *166*, 335–342. [CrossRef]
14. Ryu, C.; Shin, D. Combined Heat and Power from Municipal Solid Waste: Current Status and Issues in South Korea. *Energies* **2013**, *6*, 45–57. [CrossRef]
15. Ministerul Industriei Și Resurselor. *Normativ Tehnic Din 10 Ianuarie 2003 Privind Incinerarea Deșeurilor*; Technical Regulation of 10 January 2003 on waste incineration; Ministry of Industry and Resources: Bucharest, Romania, 2003.
16. European Parliament. *Directive 2008/98/EC of the European Parliament and of the Council of 19 November 2008 on Waste and Repealing Certain Directives*; Directive 2008/98/EC; Council of the European Union: Brussels, Belgium, 2008.
17. Zhou, Z.; Tang, Y.; Chi, Y.; Ni, M.; Buekens, A. Waste-to-energy: A review of life cycle assessment and its extension methods. *Waste Manag. Res.* **2018**, *36*, 3–16. [CrossRef] [PubMed]
18. Petrișor, A.-I.; Meiță, V. Implications of Spatial Sustainability on the Territorial Planning Framework in a Transition Country. In Proceedings of the Asian Conference on Sustainability, Energy and the Environment 2013, Official Conference Proceedings, Osaka, Japan, 6–9 June 2013; pp. 182–191.
19. Ozcan, H.K.; Guvenc, S.Y.; Guvenc, L.; Demir, G. Municipal Solid Waste Characterization According to Different Income Levels: A Case Study. *Sustain. J. Rec.* **2016**, *8*, 1044. [CrossRef]
20. Da Silva, L.J.D.V.B.; Dos Santos, I.F.S.; Mensah, J.H.R.; Gonçalves, A.T.T.; Barros, R.M. Incineration of municipal solid waste in Brazil: An analysis of the economically viable energy potential. *Renew. Energy* **2020**, *149*, 1386–1394. [CrossRef]
21. Aryampa, S.; Maheshwari, B.; Sabiiti, E.; Bateganya, N.L.; Bukenya, B. Status of Waste Management in the East African Cities: Understanding the Drivers of Waste Generation, Collection and Disposal and Their Impacts on Kampala City's Sustainability. *Sustain. J. Rec.* **2019**, *11*, 5523. [CrossRef]
22. Maramureș County Sustainable Development Strategy for the Period 2014–2020. Available online: https://www.cjmaramures.ro/attachments/strategie/Strategia%20de%20Dezvoltare%20Durabila%20a%20Judetului%20Maramures%202014-2020.pdf (accessed on 26 March 2020).
23. OpenStreetMap. Available online: https://www.openstreetmap.org (accessed on 10 March 2020).
24. Maramureș Environmental Protection Agency. 2018-County Report on the State of the Environment. 2017. Available online: http://www.anpm.ro/web/apm-maramures/rapoarte-anuale1 (accessed on 27 March 2020).
25. Maramureș Environmental Agency. 2019–County Report on the State of the Environment. 2018. Available online: http://www.anpm.ro/web/apm-maramures/rapoarte-anuale1 (accessed on 27 March 2020).
26. Menikpura, S.N.M.; Basnayake, B.F.A.; Boyagoda, P.B.; Kularathne, I.W. Estimations and Mathematical Model Predictions of Energy Contents of Municipal Solid Waste (MSW) in Kandy. *Trop. Agric. Res.* **2007**, *19*, 389–400.
27. Puiu, D.; Cruceru, L.V. Guidance tool for waste analysis in an accredited laboratory. *INCD ECOIND–International Symposium–SIMI 2016 "The Environment and the Industry"*. 2016, pp. 81–88. Available online: https://ibn.idsi.md/sites/default/files/imag_file/The%20Environment%20and%20the%20Industry_2017_0.pdf (accessed on 15 October 2020).
28. ISO (International Organization for Standardization). *Solid Mineral Fuels–Determination of Gross Calorific Value by the Bomb Calorimetric Method and Calculation of Net Calorific Value*; Czech Office for Standards, Metrology and Testing: Prague, Czech Republic, 2010; ISO 1928:2009.

MDPI
St. Alban-Anlage 66
4052 Basel
Switzerland
Tel. +41 61 683 77 34
Fax +41 61 302 89 18
www.mdpi.com

Processes Editorial Office
E-mail: processes@mdpi.com
www.mdpi.com/journal/processes